Badlands of the Republic

RGS-IBG Book Series

The Royal Geographical Society (with the Institute of British Geographers) Book Series provides a forum for scholarly monographs and edited collections of academic papers at the leading edge of research in human and physical geography. The volumes are intended to make significant contributions to the field in which they lie, and to be written in a manner accessible to the wider community of academic geographers. Some volumes will disseminate current geographical research reported at conferences or sessions convened by Research Groups of the Society. Some will be edited or authored by scholars from beyond the UK. All are designed to have an international readership and to both reflect and stimulate the best current research within geography.

The books will stand out in terms of:
- the quality of research
- their contribution to their research field
- their likelihood to stimulate other research
- being scholarly but accessible.

For series guides go to www.blackwellpublishing.com/pdf/rgsibg.pdf

Published

Badlands of the Republic: Space, Politics and Urban Policy
Mustafa Dikeç

Geomorphology of Upland Peat: Erosion, Form and Landscape Change
Martin Evans and Jeff Warburton

Spaces of Colonialism: Delhi's Urban Governmentalities
Stephen Legg

People/States/Territories
Rhys Jones

Publics and the City
Kurt Iveson

After the Three Italies: Wealth, Inequality and Industrial Change
Mick Dunford and Lidia Greco

Putting Workfare in Place
Peter Sunley, Ron Martin and
Corinne Nativel

Domicile and Diaspora
Alison Blunt

Geographies and Moralities
Edited by Roger Lee and David M. Smith

Military Geographies
Rachel Woodward

A New Deal for Transport?
Edited by Iain Docherty and Jon Shaw

Geographies of British Modernity
Edited by David Gilbert, David Matless
and Brian Short

Lost Geographies of Power
John Allen

Globalizing South China
Carolyn L. Cartier

Geomorphological Processes and Landscape Change: Britain in the Last 1000 Years
Edited by David L. Higgitt and
E. Mark Lee

Forthcoming

Politicizing Consumption: Making the Global Self in an Unequal World
Clive Barnett, Nick Clarke, Paul Cloke and
Alice Malpass

Living Through Decline: Surviving in the Places of the Post-Industrial Economy
Huw Beynon and Ray Hudson

Swept up Lives? Re-envisaging 'the Homeless City'
Paul Cloke, Sarah Johnsen and Jon May

Climate and Society in Colonial Mexico: A Study in Vulnerability
Georgina H. Endfield

Resistance, Space and Political Identities
David Featherstone

Complex Locations: Women's Geographical Work and the Canon 1850–1970
Avril Maddrell

Driving Spaces: A Cultural–Historical Geography of England's M1 Motorway
Peter Merriman

Geochemical Sediments and Landscapes
Edited by David J. Nash and
Sue J. McLaren

Mental Health and Social Space: Towards Inclusionary Geographies?
Hester Parr

Domesticating Neo-Liberalism: Social Exclusion and Spaces of Economic Practice in Post Socialism
Adrian Smith, Alison Stenning,
Alena Rochovská and Dariusz Świątek

Value Chain Struggles: Compliance and Defiance in the Plantation Districts of South India
Jeffrey Neilson and Bill Pritchard

Badlands of the Republic

Space, Politics and Urban Policy

Mustafa Dikeç

Blackwell
Publishing

BLACKWELL PUBLISHING
350 Main Street, Malden, MA 02148-5020, USA
9600 Garsington Road, Oxford OX4 2DQ, UK
550 Swanston Street, Carlton, Victoria 3053, Australia

First published 2007 by Blackwell Publishing Ltd

1   2007

Library of Congress Cataloging-in-Publication Data

Dikeç, Mustafa, 1971–
    Badlands of the republic : space, politics, and urban policy / Mustafa Dikeç.
        p.   cm.
    Includes bibliographical references and index.
    ISBN 978-1-4051-5631-8 (hardcover : alk. paper) — ISBN 978-1-4051-5630-1 (pbk. : alk. paper) 1. Urban policy—France. 2. Sociology, Urban—France. 3. Immigrants—France. 4. Identity (Psychology)—France. I. Title.

    HT135.D55 2007
    307.76086′9120944—dc22

                                                                                                    2006102118

A catalogue record for this title is available from the British Library.

The publisher's policy is to use permanent paper from mills that operate a sustainable forestry policy, and which has been manufactured from pulp processed using acid-free and elementary chlorine-free practices. Furthermore, the publisher ensures that the text paper and cover board used have met acceptable environmental accreditation standards.

For further information on
Blackwell Publishing, visit our website:
www.blackwellpublishing.com

For my son,

Aslan Hancock Dikeç

(2005)

In memoriam

Contents

List of Figures and Tables   ix
List of Abbreviations and Acronyms   xi
Series Editors' Preface   xiii
Acknowledgements   xiv

Part I   Badlands   **1**

1   Introduction: The Fear of 'the *Banlieue*'   **3**
The Colour of Fear   7
Organization of the Book   14

2   State's Statements: Urban Policy as Place-Making   **16**
Neoliberalism, Neoliberalization and the City   24
The Republican State and Its Contradictions   28
The Republican Penal State and Urban Policy   31

Part II   The Police   **35**

**3   The Right to the City? Revolts and the Initiation of
Urban Policy**   **37**
The Hot Summer of 1981: How Novel is 'Violence'?   39
Brixton in France? The Haunting of the French Republic   40
The 'Founding Texts' of Urban Policy   48
The 'Anti-immigrant Vote'   56
Consolidation of Urban Policy   60
Conclusions: Consolidation of the Police   64

**4   Justice, Police, Statistics: Surveillance of Spaces of
Intervention**   **68**
When the Margin is at the Centre   71

The 'Return of the State'   75
'I Like the State'   78
Justice, Police, Statistics   80
Conclusions: Looking for a 'Better' Police . . .   87
. . . a 'Republican' One   90

5 **From 'Neighbourhoods in Danger' to 'Dangerous**
   Neighbourhoods': The Repressive Turn in Urban Policy   93
   Encore! The Ghost Haunting the French Republic   94
   Pacte de Relance: Old Ghosts, New Spaces   97
   'They are Already Stigmatized': Affirmative Action *à la*
      française   99
   Is 'Positive Discrimination' Negative?   104
   Insecurity Wins the Left: The Villepinte Colloquium   106
   Remaking Urban Policy in Republican Terms   109
   Whither Urban Policy?   114
   The Police Order and the Police State   118
   Back to the Statist Geography   120
   Conclusions: Repressive Police   124

Part III   Justice in *Banlieues*   127

6 **A 'Thirst for Citizenship': Voices from a *Banlieue***   129
   Vaulx-en-Velin between Official Processions and Police Forces   129
   Vaulx-en-Velin after the *trente glorieuses*   131
   A 'Thirst for Citizenship'   134
   A Toil of Two Cities (in One)   136
   Whose List is More 'Communitarian'?   142
   Conclusions: Acting on the Spaces of the Police   147

7 **Voices into Noises: Revolts as Unarticulated Justice**
   Movements   152
   Revolting Geographies   153
   Geographies of Repression: 'Police Everywhere, Justice
      Nowhere'   158
   Policies of Urgency: '20 Years for Unemployment, 20 Minutes
      for Insecurity'   162
   Conclusions: From a 'Just Revolt of the Youth' to
      'Urban Violence'   166

8 **Conclusion: Space, Politics and Urban Policy**   170

   Notes   178
   References   194
   Index   212

Figures and Tables

FIGURES

1.1 and 1.2	The changing colour of 'the *banlieue*'	9
3.1	Brixton in France?	44
3.2	Regions and major cities of mainland France	45
3.3	Departments and major cities of Rhône-Alpes	45
3.4	The '3Vs' of Lyon	46
3.5	The first urban policy neighbourhoods (1982–3)	55
3.6	The 148 urban policy neighbourhoods (1984–8)	61
3.7	The 400 urban policy neighbourhoods per department (1989–93)	65
4.1	The 'difficult neighbourhoods' of the French Intelligence Service per department (1999)	84
4.2	The urban policy neighbourhoods per department (1999)	84
5.1	The location of *Maisons de Justice et du Droit* (MJDs) in mainland France in 2002	109
6.1	Social housing neighbourhoods of Vaulx-en-Velin seen from the new centre	141
6.2	The new centre of Vaulx-en-Velin seen from the social housing neighbourhoods	142

TABLES

4.1	Characteristics of the priority neighbourhoods of urban policy in comparison with their agglomerations and metropolitan France, 1990 (%)	85

5.1	Characteristics of priority neighbourhoods of urban policy in comparison with cities and metropolitan France, 1990–9 (%)	117
6.1	Unemployment rates in Vaulx-en-Velin and its urban policy neighbourhoods compared to departmental, regional and national rates for selected years (%)	132
6.2	Unemployment rates and population characteristics of Vaulx-en-Velin and its urban policy neighbour- hoods, 1990 and 1999 (%)	133
6.3	Rates of increase in unemployment levels in Vaulx-en-Velin and its urban policy neighbourhoods compared to departmental, regional and national rates for selected periods (%)	134

Abbreviations and Acronyms

ANRU	*Agence Nationale pour la Rénovation Urbaine*
CIV	*Comité Interministériel des Villes*
CLSs	*Contrats Locaux de Sécurité*
CNDSQ	*Commission Nationale pour le Développement Social des Quartiers*
CNPD	*Conseil National de Prévention de la Délinquance*
CNV	*Conseil National des Villes*
CPE	*Contrat Première Embauche*
DIV	*Délégation Interministérielle à la Ville*
DSQ	*Développement Social des Quartiers*
DSU	*Développement Social Urbain*
FN	*Front National*
GAMs	*Groupes d'Action Municipale*
GPUs	*Grands Projets Urbains*
GPVs	*Grands Projets de Ville*
HCI	*Haut Conseil à l'Intégration*
HLM	*Habitation à Loyer Modéré*
HVS	*Habitat et Vie Sociale*
INED	*Institut National d'Etudes Démographiques*
INSEE	*Institut National de la Statistique et des Etudes Economiques*
ISE	*Indice Synthétique d'Exclusion*
LOV	*Loi d'Orientation pour la Ville*
LSI	*Loi pour la Sécurité Intérieure*
LSQ	*Loi relative à la Sécurité Quotidienne*
MIB	*Mouvement de l'Immigration et des Banlieues*
MJD	*Maison de Justice et du Droit*
ORUs	*Opérations de Renouvellement Urbain*
PCF	*Parti Communiste Français*
PS	*Parti Socialiste*

RG	*Renseignements Généraux*
RMI	*Revenu Minimum d'Insertion*
RPR	*Rassemblement pour la République*
SRU	*Loi de Solidarité et Renouvellement Urbains*
UDF	*Union pour la Démocratie Française*
UMP	*Union pour un Mouvement Populaire*
ZAC	*Zone d'Aménagement Concerté*
ZEP	*Zone d'Education Prioritaire*
ZFU	*Zone Franche Urbaine*
ZRU	*Zone de Redynamisation Urbaine*
ZUP	*Zone à Urbaniser par Priorité*
ZUS	*Zone Urbaine Sensible*

Series Editors' Preface

Like its fellow RGS-IBG publications, *Area*, the *Geographical Journal* and *Transactions*, the RGS-IBG Book Series only publishes work of the highest international standing. Its emphasis is on distinctive new developments in human and physical geography, although it is also open to contributions from cognate disciplines, such as anthropology, chemistry, geology and sociology, whose interests overlap with those of geographers. The Series places strong emphasis on theoretically-informed and empirically-strong texts. Reflecting the vibrant and diverse theoretical and empirical agendas that characterize the contemporary discipline, contributions are expected to inform, challenge and stimulate the reader. Overall, the RGS-IBG Book Series seeks to promote scholarly publications that leave an intellectual mark and change the way readers think about particular issues, methods or theories.

For series guides go to www.blackwellpublishing.com/pdf/rgsibg.pdf

Kevin Ward (University of Manchester, UK) and
Joanna Bullard (Loughborough University, UK)
RGS-IBG Book Series Editors

Acknowledgements

Although none of the chapters in this book have previously appeared in their current form, I have used material from some of my published articles. These include 'Space, politics, and the political', *Environment and Planning D: Society and Space*, 2005, 23(2): 171–88 (Pion Limited, London); 'Badlands of the Republic? Revolts, the French state, and the question of *banlieues*', *Environment and Planning D: Society and Space*, 2006, 24(2): 159–63 (Pion Limited, London); 'Two decades of French urban policy: From social development of neighbourhoods to the republican penal state', *Antipode*, 2006, 38(1): 59–81 (Blackwell, Oxford); and 'Voices into noises: Ideological determination of unarticulated justice movements', *Space & Polity*, 2004, 8(2): 191–208 (http://www.tandf.co.uk/journals). I thank the publishers for permission to re-use material from these articles.

This project started under the supervision of Edward Soja, whose recommendations and encouragement are much appreciated. I am grateful to Julie-Anne Boudreau, Philippe Estèbe, Liette Gilbert, Steve Hinchliffe, Kirstie McClure, Doreen Massey, Walter Nicholls, Chris Pickvance and Michael Storper for the exciting discussions we have had over the years. Nick Henry and Kevin Ward provided constructive comments as the editors of the series at the time of writing. During the production process, it has been a pleasure to work with Angela Cohen, Rebecca du Plessis, Justin Dyer and Jacqueline Scott. Much of the work for this book was completed while I was a postdoctoral research fellow at the geography department of the Open University. During the final stages, I have enjoyed being in the wonderful working environment of the geography department at Royal Holloway, University of London. Philip Crang, Felix Driver and David Gilbert have been most supportive. Both institutions have been particularly welcoming, and their financial aid for the transcriptions and copyrights is much appreciated. John Allen, Nigel Clark, Mathew Coleman and Steve

Pile read the manuscript and provided constructive comments. I have been very lucky to have them as colleagues, and, above all, as friends.

It is hard to adequately express my gratitude to my parents and sister. And it is even harder to do so when it comes to Claire; I simply cannot find the words to express how grateful I am for her love, generosity and patience. Academic work can become very absorbing, and I definitely am not immune to this. At the early stages of this project, while I was desperately trying to finish my dissertation, Joakim joined us. He showed me how wonderful life is, and how trivial my problems were as an aspiring academic. But such lessons are easily forgotten. His brother, Aslan, arrived while I was trying to finish this book. He taught me that there is more to life than just the book – that there is life itself, with which he was not gifted. This book is dedicated to his memory.

Mustafa Dikeç
Paris and London, January 2007

Part I

Badlands

1

Introduction: The Fear of 'the *Banlieue*'

The accusations were serious: armed robbery, killing of three police officers and murder of one taxi driver. They were hurled at a young woman of 23 years old and her companion, a young man of about the same age, who was shot dead during his confrontation with the police. The evidence presented at the court, and the presence of eyewitnesses, left little hope for the young woman. The prosecuting attorney insisted on the truly cynical nature of the acts of the two, which, it was maintained, could not be justified by the circumstances. The prosecutor claimed:

> [They] are not *terrorists*, they are not Bonnie and Clyde, they are not the characters of *Natural Born Killers*. They are neither *zonards*,[1] nor *drug addicts*, nor *banlieue outcasts* [*des exclus de banlieue*]. [She] is not the daughter of *immigrants*, her mother was a teacher and helped her with homework in the evenings. These are two students who dropped out of college, gave up on work, who chose to live in a squat and to live from hold-ups, because 'money is freedom'. (*Libération*, 30 September 1998: 15; emphasis added)

What the accused were *not* associated with – terrorism, drugs, exclusion, immigration – exemplifies some of the terms that have been articulated with the spatial references of the prosecuting attorney – *zones* and *banlieues* – in the last two decades. Was the attorney, with these statements, recognizing the difficulties of growing up or living in a *zone* (being a '*zonard*') or *banlieue*? Or was she, if unwittingly, demonstrating the naturalization of crime as associated with *zones* and *banlieues*? If the accused were *zone* or *banlieue* inhabitants, would their acts be seen as more 'natural' rather than truly cynical? In a republic that cherishes so dearly the principle of equality, how can such spatial references be presented as potentially mitigating circumstances?

The attorney's argument gives us a sense of the pervasiveness of the negative image of *banlieues*, and shows how common and accepted this image has become (although there are many prestigious *banlieues* as well). This book is about a specific urban policy programme conceived to address the problems of social housing neighbourhoods in *banlieues* of French cities, which, as I will try to show, contributed largely to the consolidation of negative images associated with them. This programme was initiated by the Socialist government as an urgent response to the so-called 'hot summer' of 1981, marked by revolts in the *banlieues* of several cities. 'Urban policy', hereafter, refers to this particular policy. Conceived originally as a 'spatialization of social policies' (Chaline, 1998), it was regrouped later in 1988 under the generic term '*la politique de la Ville*' as a national urban policy with the *banlieues* as its main object. As the issues around *banlieues* have wider resonance, with connotations ranging from threats to French identity to terrorism, French urban policy, as Béhar (1999) wrote, has probably been the most debated public policy of the last two decades. This book provides a wide-ranging analysis of this policy by bringing together policy discourses and alternative voices expressed in its intervention areas. It offers an approach to urban policy that makes space central, and looks at the ways in which space is imagined and used in policy formation in the broader context of state restructuring. In so doing, it provides insight into the relationship between space and politics.

The French case is particularly important for exploring the relationship between space and politics, as space – and not community, as in the British and North American urban policy experience – has been the main object of French urban policy. This is almost necessarily so since the French republican tradition emphasizes a common culture and identity, and any reference to communities is deliberately avoided because they imply separatism, which is unacceptable under the principle of the 'one and indivisible' republic. Yet, while space remained the main object, there have been considerable changes in how space has been imagined and manipulated over the two decades of this policy. This book makes these changes and their varying political implications central to its approach to urban policy. It shows how French urban policy has constituted its spaces of intervention, associated problems with them, legitimized particular forms of state intervention, and how alternative voices formulated in such spaces challenged official designations. It situates its analysis in a broader political and economic context, showing how it feeds down into urban policy.

This book's approach to urban policy follows from a central premise to consider space not as given, but as produced through various practices of articulation. Since urban policy conceives of its object spatially, I see urban policy as a practice of articulation that constitutes space, an institutionalized practice that defines spaces (i.e. its spaces of intervention). Thus,

I maintain that urban policy constitutes its spaces of intervention as part of the policy process, rather than by acting on given spaces.

However, each policy discourse and programme is guided by particular ways of imagining space. For example, spaces of intervention may be imagined as self-contained areas with rigid boundaries, as parts of a larger network, or as part of a relational geography. Each of these ways of conceiving space has different implications for the constitution of perceived problems and the formulation of solutions to them, ranging from limited local initiatives to regional distributive policies. Thus, I insist that conceptualizations of space matter in policy, and look at the ways in which space is conceived and their policy and political implications.

Although urban policy is one way of constituting space, it is not the only one. Therefore, I bring together official discourses and alternative voices, and insist that analyses of urban policy consider policy from above and voices from below as a contestation for space. In other words, rather than merely focusing on the official discourses on *banlieues*, I try to give voice to alternative discourses formulated in *banlieues*.

My analysis, further, situates French urban policy in a wider political and economic context, and focuses on how it has constituted its spaces of intervention and how alternative voices have challenged its official descriptions. Theoretically informed by Jacques Rancière's political thought – which draws attention to the relationship between space and politics – and using Philip Corrigan and Derek Sayer's (1985) notion of 'the state's statements' – which draws attention to state's practises of articulation – I see urban policy as a particular regime of representation that consolidates a certain spatial order through descriptive names, spatial designations, categorisations, definitions, mappings and statistics. In this sense, it is a place-making practice that spatially defines areas to be treated, associates problems with them, generates a certain discourse, and proposes solutions accordingly. I do not, therefore, see urban policy as a merely administrative and technical issue, and argue against such an approach that it is tightly linked to other issues, ranging from immigration politics to economic restructuring. Instead, I adopt an eclectic approach that carries some of the features of political economy, social constructionist and governmentality approaches to urban policy. Political economy approaches relate urban policy to the larger restructurings of the state, and highlight processes of neoliberalization, premised on the extension of market relations that privilege competition, efficiency and economic success. While endorsing the attention given to the relationship between urban policy and state restructuring, I argue that there are other political rationalities that affect contemporary transformations of states and urban policy, and that equal attention should be given to established political traditions – in this case, the French republican tradition, which emphasizes the social obligations of the state towards its citizens as well as a common culture and identity, seen to be the basis of the integrity of the 'one and indivisible' republic. Such an emphasis on state restructuring and established political traditions shows that the contemporary restructuring of the French state involves an articulation of neoliberalism with the French republican tradition, producing a hybrid form of neoliberalism. It also points to the relationship between urban policy and state restructuring, which, in the French case, is manifest in the consolidation of the penal state mainly in and through the spaces of urban policy.

Although there are many parallels between the approach I adopt in this book and social constructionist and governmentality approaches, two major differences remain. First, I try to avoid the implication (usually associated with constructivist approaches; see Campbell, 1998 for a critique) that policy makers and other state actors are consciously and deliberately engaged in a discursive construction of 'reality' from a privileged place outside the domain of their very engagement, with the tools and force of language at their disposal. What interests me here is the ways in which policies put in place certain 'sensible evidences' (policy documents, spatial designations, mappings, categorizations, namings and statistics) and their effects: that is, how they help to consolidate a particular spatial order and encourage a certain way to think about it. As we will see, the kinds of sensible evidences employed, their significance and effects depend highly on the broader political and economic context; they do not, in other words, materialize in a vacuum.

Second, I argue that analyses of urban policy guided by these approaches have given insufficient attention to the issue of space (which is also observed by some scholars committed to these approaches; see, for example, Murdoch, 2004; Raco, 2003). Social constructionist approaches, while helpfully focusing on the construction of urban problems and policy discourses, neglect the role that space plays in such constructions. Governmentality approaches, on the other hand, present such an overarching argument that there is little or no room left for the difference that space makes in policy formation and resistance to it. I share the view with the social constructionist approaches that problems and policies associated with spaces of urban policy are constructed – rather than already given – but insist that equal attention be given to the ways in which such spaces are imagined and used in the formation of problems and policies. With the governmentality approaches, I concur that the construction of spaces through urban policy has a governmental dimension, but maintain that there is no inherent politics to such constructions. In other words, variations in the ways space is imagined and manipulated matter.

Approached this way, the French experience offers us the following four lessons on the nature of urban policy and on the relationship between space and politics. First, urban policy has to be understood in a range of established political traditions – in this case, French republicanism – and major national and international events – from riots in Brixton to demonstrations of high school students in Paris, from the Rushdie affair to the Islamic headscarf affair, from the Intifada to riots in Los Angeles. Second, the spaces of urban policy cannot be taken for granted, and any analysis of urban policy has to critically analyse the ways in which policies constitute their spaces of intervention. Third, ways of imagining space influence both the definition of problems associated with intervention areas and policy responses to them. In more general terms, different ways of imagining space have different political implications. Finally, both governance and resistance are spatial, place-making practices. In this sense, there is an ongoing contestation for space: what the official policy discourse constitutes as 'badlands' also become sites and organizing principles of political mobilization with democratic ideals.

The 'badlands' in question are the *banlieues* of French cities: that is, neighbourhoods in the peripheral areas of cities. In order to understand what is at stake in French urban policy, we need first to get a sense of what the *banlieues* stand for.

The Colour of Fear

Banlieue literally means suburb, but it carries different connotations from the ones associated with the British or North American suburb. Originally an administrative concept, the term *banlieue* geographically denotes peripheral areas of cities in general.[2] Such a geographical designation is not necessarily negative (as in 'the *banlieue*'). Nevertheless, the term evokes an image of excluded places, as its etymological origin suggests:

> '*Ban*' comes from the earliest medieval times, when it meant both the power of command and the power of exclusion as part of the power of command. Banned [*Banni*], banishment [*banissement*], *banlieue* – all these terms have the same origin; they refer to places of exclusion. Clearly, *banlieues* have existed independently from terms to designate them, they have made and often managed their own history, they have not simply been excluded places, but their existence does nevertheless express this will to create on the outskirts of the city places that do not belong to the system. (Paul-Levy in Banlieues 89, 1986: 125)[3]

Now the term mostly evokes an image of a peripheral area with concentrations of large-scale, mostly high-rise social housing projects, and problems associated, in the US and the UK, with inner-city areas. It no longer serves merely as a geographical reference or an administrative concept, but stands for alterity, insecurity and deprivation. In order to emphasize the term's origin and geographical connotations, I use *'banlieue'* instead of 'suburb' throughout the text.

In the early 1980s, Rey (1999: 274) writes, the *banlieues* of large French cities began to 'arouse a feeling of fear', a feeling that continued to increase in the decades to follow, becoming one of the major 'phobias' of the French in the new millennium (*Libération*, 8 April 2002: 4–5). The term *'banlieue'* designates the social housing estates of popular neighbourhoods in the peripheral areas of cities as threats to security, social order and peace. This threat, furthermore, has become closely associated with the populations living in *banlieues*, often defined in 'ethnic' terms. The fear of the *banlieue* is closely associated with a feeling of insecurity and a fear of immigration (Rey, 1999).

A similar observation is made by Hargreaves, who argues that the 1990s was a turning point in the eventual association of the *banlieue* with a feeling of insecurity and a fear of immigration:

> During the 1990s, *a new social space has been delineated* in France: that of the *'banlieue(s)'* (literally, 'suburb(s)'). A term that once served simply to denote peripheral parts of urban areas has become a synonym of alterity, deviance, and disadvantage. The mass media have played a central role in this re-construction, in the course of which they have disseminated and reinforced stereotypical ideas of people of immigrant origin as fundamentally menacing to the established social order. (Hargreaves, 1996: 607; emphasis added)

Hargreaves exemplifies the media creation of 'the *banlieues* as a news category' and the amalgamation of 'urban deprivation, immigration, and social order' in the 1990s with an issue of the journal *L'Express*, which presented a cover story under the title *'Banlieues* – Immigration: State of Emergency' (5–12 June 1991). The same journal, however, had presented another similar cover story almost two decades earlier under the title *'Banlieues*: "Hooligans" are Talking to You'. The subtitle read: 'At the gates of large cities, thousands of hoodlums are produced' (3–9 September 1973). As the cover drawing and the photos depicted them, the hooligans and hoodlums of *L'Express* in 1973 were all white. They would change colour in 1991, but the spatial reference would remain the same. In this sense, *L'Express* best exemplifies the changing colour of the fear of 'the *banlieue*' from the 1970s into 1990s (Figures 1.1 and 1.2).

Media reviews provide clear examples of the changing image of the *banlieues* in the last two decades (see, for example, Collovald, 2000, 2001; Hargreaves, 1996; Macé, 2002). However, the current image of the *banlieues* is not simply the product of journalistic accounts. Many of the

1.1

1.2

Figures 1.1 and 1.2 The changing colour of 'the *banlieue*' (1.1 (*head*): '*Banlieues*: "Hooligans" are Talking to You'. *Source: L'Express*, 3–9 September, 1973; 1.2 (*foot*): '*Banlieues* – Immigration: State of Emergency'. *Source: L'Express*, 5–12 June, 1991) journalistic categories used to frame *banlieues* have been *institutionalized* by state policies. The period in which the *banlieues* became articulated with issues of immigration, insecurity and social order was a period of intense official engagement with the question of *banlieues* – notably through urban policy, which became increasingly concerned with issues of immigration and insecurity, often to the detriment of its initial social and democratic ideals. It is these changes that I will chart in the following chapters, placing them in broader political and economic context, and relating them to the contemporary restructuring of the French state along increasingly authoritarian and exclusionary lines.

As I will try to show, the contemporary restructuring of the French state is marked by a strong attachment to the republican tradition. The French conception of the republic emphasizes a common culture and identity, fragmentation of which is seen as a threat to the social and political integrity of France. The republican tradition is based on the presupposition that 'without a common culture and a sense of common identity, the political as well as physical integrity of France would be "threatened"' (Jennings, 2000: 586). There is, therefore, little or no room for claims rising from 'differences'. The French citizen is a universal individual-citizen, directly linked to the nation-state, and national-political membership requires the acceptance of French cultural values (Feldblum, 1999; Safran, 1990). There is no official recognition of ethnicity, race or religion as intermediary means for obtaining particular rights, and the very notion of minority is strongly rejected (de Rudder and Poiret, 1999). Such a conception, in the context of fascinating diversity, generates a firm suspicion towards all kinds of particularisms. As Jennings argues,

> [T]he political project of nation-building pursued by the French state led not only to a weak conception of civil society but also to the persistent fear of the dangers of 'communities' operating within the public sphere. Within this project, citizenship was grounded upon a set of democratic political institutions rather than upon a recognition of cultural and/or ethnic diversity. Republicanism itself thus became a vehicle of both inclusion *and* exclusion. (2000: 597)

It is from this deep attachment to the republican tradition that follows what Hargreaves (1997: 180) calls a republican myth of the French nation characterized by an 'apparent blindness or outright hostility to cultural diversity', which not only leaves little or no room to cultural 'differences' (Wieviorka, 1998), but also enhances 'a system of intimidation that interdicts all protest social movements on the part of minority groups, without providing them the means to fight against inequalities and oppression of which they remain the victims' (de Rudder and Poiret, 1999: 398–99).

Such a concern with French identity and cultural differences was perhaps best exemplified by a 1992 report of the *Haut Conseil à l'Intégration* (HCI), a council created in 1989 to advise the government on the issue of integration based on a 'republican model'.

Notions of a 'multicultural society' and the 'right to be different' are unacceptably ambiguous. It is true that the concept of the nation as a cultural community [. . .] does appear unusually open to outsiders, since it regards an act of voluntary commitment to a set of values as all that is necessary. But it would be wrong to let anyone think that different cultures can be allowed to become fully developed in France. (HCI, 1992; cited in Hargreaves, 1997: 184)

It should be noted, however, that there has been a renewed enthusiasm for the republican tradition with nationalistic overtones since the 1990s, which I refer to as 'republican nationalism'. The rise of republican nationalism has been observed by many scholars with regard, in particular, to citizenship and immigration policies (see, for example, Balibar, 2001; Blatt, 1997; Feldblum, 1999; Tévanian and Tissot, 1998). As we will see, urban policy has also been influenced by the development and deployment of republican nationalism since the 1990s.

Before moving on, a preliminary explanation of certain notions might be helpful as the republican tradition shapes political debate and policy discourse in a particular way. We will see that the following four notions are commonly used in policy discourse and debates around the *banlieues*: 'communitarianism', 'ghetto', 'social mixity' and 'positive discrimination'. These notions may sound ordinary and their meanings self-evident (except, probably, the last one), but they connote particular issues and carry remarkable political weight in the French context, where a common culture and identity is emphasized as a basic republican presupposition.

'Communitarianism' (*communautarisme*) is basically used to refer to 'ethnic' communities, formation of which is seen as a threat to the cultural and political integrity of the republic. It implies 'ethnic' separatism (Hargreaves, 1995). Ghettos are the spatially reified forms of this 'ethnic' nightmare haunting the republic. The term is often used in the media and by politicians, notably from the 1990s on, to refer to the deprived areas in the *banlieues*, comparing them to inner-city areas in the USA. Wacquant challenged this use in a series of articles (1992, 1993, 1995, 1999; see also de Rudder, 1992 and Hargreaves, 1995), and insisted that the comparison is highly misleading. The areas referred to as ghettos in France, he argued, are neither ethnically homogeneous nor large enough to function as self-contained areas apart from the central city. Furthermore, unemployment, poverty and crime rates are less severe compared to the ghettos in the USA. Similarly, de Rudder (1992) argued that the neighbourhoods referred to as 'ghettos' are neither institutionalized nor homogeneous, that immigrants still remain a minority, and even when there are concentrations of immigrants, they are not homogeneous in terms of their origins: 'The use of the term "ghetto" here seems to have a more ideological than descriptive function. The word causes fear [. . .] among French natives as much as immigrants themselves. Thus both exclusion and social control (or even policing) over minorities are confirmed and even justified' (de Rudder, 1992: 261). A similar argument is advanced by Hargreaves: 'Minority groups are over-represented in disadvantaged parts of French cities, but it is empirically misleading and ideologically dangerous to speak of these areas as "ghettos" ' (1995: 76). The 'ethnic' connotations of the term makes it politically significant in the French context, and, as we will see, it is widely used in policy discourse, notably from 1990 onwards.

The notion of 'social mixity' was first introduced during the debates around a law passed in 1991. Commonly referred to as the LOV (*Loi d'Orientation pour la Ville*), or as the 'anti-ghetto law', this law was aimed at a better distribution of social housing (reviewed in Chapter 4). There exists no official definition of this term, but the idea behind it is to prevent concentrations of 'ethnic' groups in social housing neighbourhoods. The term 'ethnic' is never used, since the republican principles do not allow such references.

The last notion that needs some clarification is 'positive discrimination', which may be seen as affirmative action *à la française*. In an article on French republicanism, Jennings writes that 'there remains an unshakeable insistence upon the secularism of the state and the refusal to recognize groups of persons. Only individuals exist in the eyes of the republic. There can be no possibility of a policy of "positive discrimination", precisely because it will contribute towards the "constitution of structured communities" ' (2000: 583). Yet there exists a policy of positive discrimination in France, officially recognized in an urban policy programme of 1996 (the *Pacte de Relance*, reviewed in Chapter 5). Even before that, starting with the Educational Priority Areas (ZEPs, *Zone d'Education Prioritaire*) introduced by the Socialist government in 1981, there were *spatially designated areas* that were subject to differential treatment (i.e. to 'positive discrimination'). The 1996 programme officially used the term, but added the adjective 'territorial'. This, again, has to do with the republican tradition. As Jennings (2000) notes, positive discrimination based on ethnic, cultural and religious groups is not possible under the republic. Positive discrimination, then, is only made possible by a spatial approach, which does not, explicitly at least, discriminate on the basis of ethnic origins or cultural specificities.

In addition to these notions, it would also be helpful to clarify some of the more specific terms that are commonly used in policy discourse and the media – 'zone', 'HLM', 'cité' and 'quartier'. What these terms have in common is that they are all negatively connoted spatial designations (remember the prosecuting attorney's reference to 'zones' at the outset of this section), although they do not *necessarily* carry a negative meaning. 'Zone' ('la zone', literally 'zone, area') was originally used to refer to the makeshift dwellings set up around the fortifications of Paris, and its meaning extended to refer to deprived peripheral areas. The Robert & Collins dictionary translates the term as 'slum belt'. The term is still used in a common expression, 'c'est la zone', to refer to areas perceived as remote and/or undesirable.

The term 'zone' was also used in the post-war period in order to designate areas for urbanization at the peripheral areas of cities, starting in 1958 with the Priority Urbanization Areas (ZUPs, *Zones à Urbaniser par Priorité*). Since then, many 'zones' have been designated for policy purposes, as we will see in the following chapters. For the moment, however, it may be useful to keep in mind the distinction between these two uses of the term 'zone'. The first one (as in 'c'est la zone') carries negative connotations, which has to do with the term's historical usage to refer to deprived areas outside the city. The second one is used in the policy discourse (as in ZUPs) to designate areas of intervention. Although this second usage is not necessarily negative, the two 'zones', the name and the adjective, usually overlap.

The 'HLM' (*Habitation à Loyer Modéré*, Moderate Rent Housing) is French social housing. Although the dominant image of the HLM is one of large-scale, high-rise housing development in the peripheral areas of cities, not all the HLMs conform to this image. There are HLMs that are not large-scale and high-rise, located in the central areas of cities. The dominant image of the HLM follows from the post-war urbanization pattern of rapid and mass construction in the peripheral areas of cities where land was available and cheaper.

The same is true also for 'cités', which evoke an image similar to the stereotypical HLM. A *cité* is a group of buildings constructed according to a single plan, often isolated from (or at least clearly demarcated from) the rest of the urban fabric. A *cité* might be a *cité ouvrière* (similar to company towns), *cité universitaire* (halls of residence), *cité-jardin* (garden city) or *cité-dortoir* (dormitory town). The original meaning of the term had to do with the enclosed medieval cities, and some of the old city centres are still referred to as *cités* (for example, *l'île de la Cité* in Paris, *la Cité de Carcassonne*, *la Cité de Londres*). The term also has political connotations. When, for example, one talks about 'the life of the *Cité*' (*la vie de la Cité*), the reference is to the city as a political entity, implying its political management and public life. When used in this sense, the first letter is usually, but not necessarily, capitalized. The notion of '*droit de cité*' also derives from the political implications of *cité*, and means 'right of abode'. However, the term *cité* is commonly used, to cite Wacquant's (1993: 367) definition, to refer to 'degraded working-class neighbourhoods harbouring large low-income housing tracts'. Hargreaves (1996) translates it as 'estates' or 'council estates'. I use the term without translating throughout the text as a reminder of its political implications.

Finally, '*quartier*', which is sometimes translated as 'quarter', as in 'the Latin Quarter'. *Quartier* literally means 'neighbourhood, district, area'. It could, therefore, be anywhere, in the centre of Paris, for example (the celebrated Latin Quarter is also a '*quartier*'). The term, however, acquires a particular meaning in urban policy discourse, and designates the urban policy neighbourhoods, which, mainly, are social housing neighbourhoods in the *banlieues*. Although I translate it as 'neighbourhood', it should be noted that in policy discourse, political debates and in the media, the generic term '*quartier*' refers to *certain* neighbourhoods (i.e. social housing neighbourhoods, mainly in the *banlieues*), and it conveys a negative image.[4]

To reiterate, my main point here is that French urban policy has to be seen in relation to the republican tradition, which informs its formulation of perceived problems, proposals for solutions and legitimations of state intervention. As we will see, this relationship has become more marked since the 1990s with the rise of republican nationalism, leading to the articulation of *banlieues* in increasingly 'ethnic' terms, as incompatible with – even 'threatening' – the integrity of the republic. Despite the republican anxiety over division and disunity, French urban policy operated with a more divisive spatiality, consolidating a rather rigid geography of 'threat'. This orientation also signalled the coming of the penal state, and largely undermined citizenship and justice movements in the *banlieues*.

Organization of the Book

The book is organised in three parts: (I) 'Badlands'; (II) 'The Police'; and (III) 'Justice in *Banlieues*'. Chapters 1 and 2 in Part I set the stage and propose an approach to urban policy with a focus on the spatial conceptualizations of intervention areas, and their varying political implications.

In the early 1980s, French urban policy was conceived with such stated ideals as 'self-management', 'social development of neighbourhoods', 'democratic management of the city' and 'the right to the city'. Since then, however, not only has urban policy's articulation of *banlieues* changed, but also other state institutions such as the Ministry of Justice and the French

Intelligence Service have become involved with the issue. In the process, earlier ideals have been overshadowed by increasingly authoritarian measures towards *banlieues* with a stated aim to 're-conquer no-go areas'. Chapters 3 to 5 in Part II – 'The Police' – demonstrate the transformation of urban policy from a more socially oriented policy to one obsessed with security. This transformation is placed in wider political and economic context, both national and international. Organized around three themes – 'revolts', 'surveillance' and 'repression' – that correspond to three periods (1981–9, 1990–2 and 1993–2006), chapters in this part show that while urban policy has been dealing with practically the same areas for years, the ways in which it conceived its spaces of intervention, associated problems with them, and legitimized particular forms of intervention have changed considerably.

Chapters 6 and 7 constitute Part III – 'Justice in *Banlieues*' – which takes its name from a resistance movement (as we will see in Chapter 4) that seeks to federate separate political mobilizations in *banlieues*. In Chapter 6, I tell the story of a notorious *banlieue*, Vaulx-en-Velin in the Lyon metropolitan area, which has been included in urban policy programmes since 1984. This *banlieue* was the site of furious revolts in 1990 (and later in 1992 and 2005, though of a smaller scale), and remains a major reference in debates around urban policy and *banlieues*. Through interviews with local officials and the members of a local political association founded by immigrant youth, I show that despite their negative stereotypical image as badlands, *banlieues* are also sites of political mobilization – or of 'insurgent citizenship', to use Holston's (1998) notion[5] – with democratic aspirations, drawing on a vocabulary of justice, citizenship and equality. This account is part of alternative voices that I insist should be taken into consideration in debates around urban policy. Chapter 7 presents an analysis of recurrent revolts in the *banlieues*, and shows that revolts are reactions to persistent problems such as mass unemployment, discrimination, racism and police violence, although their official framing (as explored in Part II) highlights less the difficult material conditions in the *banlieues* than the 'threat' posed by them to security and social order. Chapter 8 concludes the book by re-visiting the arguments laid out in the first chapter about space, politics and urban policy.

2

State's Statements: Urban Policy as Place-Making

States, if the pun be forgiven, *state*; the arcane rituals of a court of law, the formulae of royal assent to an Act of Parliament, visits of school inspectors, are all statements. They define, in great detail, acceptable forms and images of social activity and individual and collective identity; they regulate, in empirically specifiable ways, much – very much, by the twentieth century – of social life. Indeed, in this sense 'the State' never stops talking.

Corrigan and Sayer, 1985: 3

The power of definition, as Hobbes knew, is a highly strategic power invested in the state.

Connolly, 1991: 207

I take Corrigan and Sayer's notion of 'state's statements' as a starting point from which to work out my approach to urban policy, which, as mentioned in the previous chapter, follows from a premise to consider its spaces not as given, but as produced through practices of articulation. The notion allows me to capture the state's diverse practices of articulation as they relate to the spaces of urban policy – definitions, categorizations, spatial designations, namings, mappings and statistics. It thus enables me to highlight various articulatory practices at work – including urban policy as well as other state's statements – in the constitution of the spaces of urban policy (i.e. its spaces of intervention).

The notion also allows me to draw attention to the governmental dimension of such articulatory practices. Corrigan and Sayer's formulation implies that the state's statements are practices of articulation with a governmental dimension. How, then, could the state's statements be conceptualized in spatial terms? If the state's statements define the 'proper place' of things and people, as Corrigan and Sayer suggest, what are the political implications of such an ordering? And what are the implications of this when considering the relationship between space, politics and urban policy? Here I draw on Rancière's political thought, which informs the theoretical orientation of this book. Rancière's political writings seem to be an odd place to seek tools for looking at urban policy. However, two remarks by activists in a notorious *banlieue* called Vaulx-en-Velin led me to this endeavour.[1] The first one is by Pierre-Didier Tchétché-Apéa, complaining about the negative image of *banlieues* and increasingly repressive measures devised to contain them:

> Are we going to go on this way with millions of citizens excluded from the system, and just make do with it, with impressive police forces, even the military, to contain these areas and keep these territories *well delimited in public space*, and where serious crises like riots and so on are not considered as political acts, in the sense that they result from a socio-political problem, but are seen only in terms of security and repression? (Interview, Pierre-Didier Tchétché-Apéa, 2002; emphasis added)

The second remark is made by Yves Mena, commenting on the difficulties of political mobilization in the *banlieues*:

> *When there's an initiative in a banlieue*, there are people who in other respects you would never suspect, people who are, I don't know, members of anti-racist groups, part of movements for the defence of human rights, and who say to us, 'what you're doing is good, but it's a pity because', roughly speaking, 'there are too many foreigners in it', or else, '*you're a bit too connoted*' . . . It's dismaying to hear things like that! (Interview, Yves Mena, 2002; emphasis added)

What I retained from these remarks (see italics above) was that the status of *banlieues* as well-delimited spaces with negative connotations undermined the possibilities of opening up spaces of politics in such areas. The place assigned to them in public space not only debilitated the political significance of revolts (which was not the case when the first series of revolts took place in the early 1980s), but also undermined more conventional forms of political mobilizations. In the consolidated spatial order, *banlieues* had their 'proper place', which resonated not with politics but with pointless rioting and increasingly repressive measures.

Thinking conceptually about these remarks led me to Rancière's work as a basis for my analysis of urban policy. Rancière's main political concern is to *resist the givenness of place*. Or, as Robson (2005: 5) has succinctly put it, central to Rancière's political thought is '[a] refusal to observe the "place" allocated to people and things (or at least, to particular people and things, such as the worker or the voice of the crowd)'. The contingency of any established order of governance with its distributions of functions, people and places is the central premise of Rancière's politics.

> In the end, everything in politics turns on the distribution of spaces. What are these places? How do they function? Why are they there? Who can occupy them? For me, political action always acts upon the social as the litigious distribution of places and roles. It is always a matter of knowing who is qualified to say what a particular place is and what is done in it. (Rancière, 2003a: 201)

There exists a substantial spatial dimension in Rancière's theorization of politics (Dikeç, 2005), which implies a disruption of the spatial distributions of the established order. For Rancière, orders of governance constitute, normalize and depend upon particular spatial organizations, which become the 'naturally given' basis for government. This basis endows some the authority to govern while leaving others as the governed or the dominated. The established spatial organization provides the given on the basis of which problems are defined, solutions proposed. Furthermore, the established spatial organization, once normalized, provides a particular *locus of enunciation* for the governed, as implied by the remarks of Pierre-Didier and Yves Mena above.

The name Rancière gives to such spatial orderings is '*the police*' – understood in its original, non-pejorative sense as 'government, organization'. 'The police' refers to an established form of governance with everyone in their 'proper place' in the seemingly natural order of things. Three points should be emphasized here. First, Rancière uses the term 'the police' in a broad sense to refer to 'all the activities which create order by distributing places, names, functions' (1994: 173). The plurality of activities in his definition is important; otherwise the police would be merely a shorthand for totalitarianism. Second, the police – any police – order is contested and full of tension, and although the police *notion* of the society is based on a principle of saturation (determined spaces with everyone and everything in their 'proper' place), there is never a total closure. Third, the police is not identical to 'state apparatus'. Although state's practices of articulation may be seen as one example of the constitution of a police order, the 'distribution of places and roles that defines a police regime stems as much from the assumed spontaneity of social relations as from the rigidity of state functions' (Rancière, 1999: 29).

The police is based on a particular regime of representation, to which Rancière refers to as '*the partition of the sensible*', defined as 'that system of sensible evidences that discloses at once the existence of a common [i.e. the whole to be governed] and the partitions that define the respective places and parts in it' (2000a: 12). The partition of the sensible, as a system of sensible evidences, arranges the perceptive givens of a situation – what is in or out, central or peripheral, audible or inaudible, visible or invisible. The police, then, is not self-evident or naturally given, but rather a product of a particular regime of representation, or what Rancière calls sensible evidences. It is exemplary in this sense that one of the first measures the French Minister of the Interior Nicolas Sarkozy had proposed, when he took office in 2002 with a stated aim to 'restore the republican order', was to modify the periodicity of the publication of figures of delinquency, and to make them publicly available more frequently (*Le Monde*, 31 May 2002a).

The police, therefore, is both a principle of distribution and an apparatus of administration, which relies on a symbolically constituted organization of social space, an organization that becomes the basis of and for governance. Thus, the essence of the police is not repression but *distribution* – distribution of places, people, names, functions, authorities, activities, and so on – and the normalization of this distribution. This is where Rancière's work at once converges with and diverges from Foucault's. The convergence – and Rancière's intellectual debt to Foucault[2] – becomes clear in the following extracts:

> The notion of police, even in France today, is frequently misunderstood. When one speaks to a Frenchman about police, he can only think of people in uniform or in the secret service. In the seventeenth and eighteenth centuries, 'police' signified a program of government rationality. This can be characterized as a project to create a system of regulation of the general conduct of individuals whereby everything would be controlled to the point of self-sustenance, without the need for intervention. (Foucault, 1984: 241)

> This term ['the police'] no doubt poses a few problems. The word *police* normally evokes what is known as the petty police, the truncheon blows of the forces of law and order and the inquisitions of the secret police. But this narrow definition may be deemed contingent. Michel Foucault has shown that, as a mode of government, the police described by writers of the seventeenth and eighteenth centuries covered everything relating to 'man' and his 'happiness'. The petty police is just a particular form of a *more general order that arranges that tangible reality in which bodies are distributed in community*. (Rancière, 1999: 28; emphasis added)

For Foucault, the police extended well beyond a narrowly defined institution. Rancière shares this view, but emphasizes *distribution* as the essence of the police, and distinguishes his approach from that of Foucault's in following terms:

> In *Omnes et singulatim* Foucault treats the police as an institutional device [*dispositif*] which partakes of the control of power on life and the bodies.

> Police, in my work, does not refer to an institution of power, but to a *principle of the partition of the sensible* within which strategies and techniques of power can be defined. (Rancière, 2000b; emphasis added)

The distinctive spatiality of Rancière's political thought makes it more compelling for me in this project about the relationship between space, policy and politics. While both Foucault and Rancière are sensitive to the spatiality and contingency of orders of governance, their work has varying implications for thinking about the relationship between space and politics. Whereas, in Foucault's work, space mainly – though not exclusively – appears as part of a disciplinary technology and as 'fundamental in any exercise of power' (1984: 252), for Rancière, it forms the basis of his conceptualization of politics, which is not centred on the notion of power. In other words, while Foucault's sensibility towards the spatiality of order leads him to an exploration of the relationship between power and space (as evidenced, for example, in his discussion of the plague-stricken town and the Panopticon in *Discipline and Punish*), Rancière's results in a distinctively spatial conceptualization of politics:

> I address spatial categories as categories of distribution: distribution of places, boundaries of what is in or out, central or peripheral, visible or invisible. They are related to what I call *the partition of the sensible*. [. . .] In other words, my concern with 'space' is the same as my concern with 'aesthetics'. [. . .] My work on politics was an attempt to show politics as an 'aesthetic affair' because politics is not the exercise of power or the struggle for power. It is the configuration of a specific world, a specific form of experience in which some things appear to be political objects, some questions political issues or argumentations, and some agents political subjects. (Rancière, 2003b: 5–6)

Hence the difference between their notions of the police – as part of an institution of power in Foucault's case, and as a principle of spatial ordering in Rancière's. The police, in the way Rancière conceives it, is inherently spatial. Space is pertinent to it because identificatory distribution (naming, fixing in space, defining a proper place) is an essential component of government (as also suggested by Foucault, 1977). Such distributions define legitimate interlocutors, make sensible certain issues while making others imperceptible, distinguish voices from noises. Rancière's major political premise is that the givens of the police, its ordered spaces and partitions of the sensible, are always polemical and never objective, which is the possibility of politics: 'Political activity is whatever shifts a body from the place assigned to it or changes a place's function.[3] It makes visible what had no business being seen, and makes heard a discourse where once there was only place for noise; it makes understood as discourse what was once only heard as noise' (Rancière, 1999: 30). Politics, then, 'acts on the police' (Rancière, 1999: 33). The spaces of the police and spaces of politics are enmeshed. Politics is thus inherently spatial as it puts into question the very distributions of the police, its partitioned spaces, which are normalized by regimes of governance. It is about the givens – always polemical and never objective – of a situation, about the established order of things, including established practices of identification.

Rancière's theorization of the police and politics is important for me as it draws attention to the relationship between space, politics and (urban) policy, understood as a practice of articulation that involves *spatial ordering* through descriptive names, categorisations, definitions, designations and mappings. Following Rancière's definition of the police, I see urban policy as a particular regime of representation that consolidates a certain spatial order through such practices of articulation – spatially designating areas to be treated, associating problems with them, and generating a certain discourse (though not the only one) about them. In this sense, it is a place-making practice with material effects in everyday lives of people, as we will see with the story of Vaulx-en-Velin in Chapter 6. Adopting such an approach has three implications for looking at urban policy.

First, rather than insisting on the self-evident quality of *banlieues* as given objects of intervention, I look at the ways in which they were constituted as objects of intervention with an associated discourse that carries the authority of state's statements. To put it in Rancière's terms, I look at the sensible evidences that were put into place in the constitution of *banlieues* as objects of intervention. These include policy documents, spatial designations, mappings, categorizations, namings and statistics, all of which contributed to the consolidation of a certain spatial order and a particular way to think about it. As we will see in Part II ('The Police'), the spatial order that urban policy helped to consolidate with its designations of intervention areas became officially so accepted that when the French Intelligence Service decided to engage with the question of *banlieues*, it was the list of urban policy neighbourhoods that they took as a starting point. When the Ministry of Justice engaged with the issue with a stated aim to restore the law, its measures aimed at the same neighbourhoods. Similarly, other repressive measures (like security contracts, Sarkozy's flash-ball guns, etc.) and growing anxieties about the 'values of the republic' were all guided by the same spatial imaginary, which became the basis for the consolidation of what I call the republican penal state from the 1990s onwards.

Second, I look at the political implications of the consolidation of this spatial order. I do so by looking at the French state's responses to recurrent revolts in the *banlieues*, and relating them to the changing articulations of *banlieues* in increasingly negative terms. This shows that the political significance of revolts fades away as the *banlieues* are articulated more as a form of menacing exteriority and as a more repressive police order is consolidated. I try to show that this articulation highlights less the difficult material conditions in *banlieues* than the 'threat' posed by these areas, shifting focus from growing inequalities and discriminations to menaces to 'the values of the republic', French identity and the authority of the state.

Third, I look at *banlieues* not merely as 'badlands', as the official discourses have increasingly articulated them, but also as sites of political mobilization with democratic ideals – which is exemplified in Chapter 6. There is, in this sense, a constant tension between the official discourse that defines *banlieues* and alternative voices formulated in the areas themselves, which question the place assigned to them in the police order. What the state's statements constitute as badlands, I show, are also sites of political mobilization that are aimed at opening up political spaces in the determined spaces of the police.

However, an approach to urban policy focusing merely on the three points outlined above neglects one aspect that is distinctive about urban policy – space. As Cochrane (2000: 532) notes, urban policy is distinctive compared to other social policies in that it conceives of its object spatially:

> Urban policy [. . .] focuses on places and spatially delimited areas or the groups of people associated with them. Its problem definition starts from area rather than individual or even social group, although, of course, a concern with an area is often used as a coded way of referring to a concern about the particular groups which are believed to be concentrated in it.

This suggests that due attention be given to the ways in which space is conceived in urban policy. At the outset of this chapter, I have argued, following Corrigan and Sayer, that urban policy – as a state's statement – has a governmental dimension. I have argued, furthermore, that it is a place-making practice, constituting its spaces of intervention as part of the policy process rather than intervening in already given spaces. Yet, this is not to imply that constitution of spaces through urban policy is a governmental practice in a univocal sense. Let me elaborate.

One of the issues Foucault's (1991) notion of governmentality raises is the mutual constitution of objects of governance and modes of thought – mentality – which then makes specific forms of intervention possible. Thus, the relevance of the concept for studies of (urban) policy derives from its emphasis on the mutual constitution of formation and intervention as part of the activity of governance, of which policy-making is one aspect. The emphasis on the mutual constitution of specific forms of representation and intervention is indicated by the semantic linking of 'governing' and 'mentality' (Lemke, 2001). This implies that governmental practices cannot be considered independently of the formation of the objects and subjects of governance. As Thrift (1999: 276) notes, following Rose (1996), one of the issues Foucault intended to signify with the notion of governmentality was the ways in which governmental activities were based on particular forms of political reasoning as conceptions of the objects and subjects to be governed.[4]

This formulation has useful implications for analysis of urban policy, which depends on and deploys particular spatial imaginaries. Urban policy implies 'a process of actual boundary shaping', and may be seen as an institutionalized spatial arrangement (Shapiro and Neubauer, 1989: 303) – what I referred to above, following Rancière, as 'the police'. This makes it difficult to talk unproblematically about the 'spaces' of urban policy, because, in this view, they are constituted as part of the policy process. It follows that governmental practices, insofar as they involve both formation and intervention, are not merely 'confined' to designated spaces; they constitute those spaces as part of the governing activity. If urban policy has a governmental dimension, as I have suggested, then its spaces of intervention are not merely the sites of this governmental practice, but, first and foremost, its outcomes. However, policy discourses and programmes are guided by particular ways of imagining space, which have different implications for the constitution of problems and formulation of solutions. In other words, how space is conceived matters because urban policy conceives of its object spatially. How do different urban policies imagine their spaces of intervention? What kind of a spatial imaginary do they constitute and act upon? What difference do different ways of conceiving space make for urban policy? Such questions have received little – if any – attention in the geographical literature that focuses on governmentality and urban policy.

The French case, however, suggests that they should. I will show in Part II that although French urban policy has been addressing practically the same areas for years, the ways in which its intervention areas have been spatially conceptualized (and discursively articulated) have changed remarkably. This is not to imply that each succeeding programme has marked a rupture with the preceding one. In each of the three periods I use to organize my account, however, there was a discernible change not only in forms of state intervention – from revolts and the initiation of urban policy (1981–9) to surveillance (1990–2), and eventually to repression (1993–2006) – but also in the spatial conceptualizations of intervention areas – from a relational view of intervention areas to self-contained areas with rigid boundaries. Different ways of imagining space, I will argue, had different implications for the constitution of problems and formulation of solutions. Furthermore, each of the changing conceptualizations followed different political rationalities, understood as conceptualizations and justifications of goals, ideals and principles of government (Rose and Miller, 1992; Simons, 1995). This implies that the constitution of spaces through urban policy has a governmental dimension, but there is no inherent politics to such constitutions. It seems to me important, therefore, to give adequate attention to varying conceptualizations of space as they have different political implications, and reflect different political rationalities.

This leads me to the issue of political rationalities. State's statements are not produced in a vacuum. They change as governments change and states restructure, they reflect or deflect certain ideologies, follow particular rationalities, and respond to events, both local and global. In other words, the political-economic context in which they are produced matters. This issue has been effectively dealt with in the literature on urban policy and state restructuring, which interprets the contemporary transformations of western states and urban policies as a process of neoliberalization. It is to these debates that I now turn, which provide useful insights for interpreting urban policy in the larger framework of state restructuring.

Neoliberalism, Neoliberalization and the City

One of the key elements of the burgeoning geographical literature on neoliberalism is its emphasis on the production of new spaces and regulatory frameworks in interpreting contemporary transformations of cities and states. Such an emphasis encourages an interpretation of neoliberalism not as something merely happening to cities and states, but rather as a specific form of political rationality producing new spaces, and at the same time produced upon and through the spaces of *particular* states and cities. An important implication of such an interpretation is that it allows for variation: if neoliberalism is itself produced upon and through diverse geopolitical spaces (and also producing them), then there should be hybrid forms of neoliberalism (Larner, 2003; Peck, 2004). Yet such variations have not figured in the accounts of neoliberalism and neoliberalization in geographical literature, which are mainly inspired by US, UK and, to some extent, Canadian cities and states. Although the perils of reproducing one hegemonic story of neoliberalism are recognized and the hybrid nature of neoliberalism is emphasized, little or no empirical evidence is provided.

'Despite the familiarity of the term, defining neoliberalism is no straightforward task,' write McCarthy and Prudham (2001: 276). This is so, they argue, partly because the term refers to a variety of principles and practices, too complex to be effectively captured by a single definition. There is also concern about the homogenization of specific forms of neoliberalism under overarching definitions (Larner, 2003; Peck, 2004). Defining neo-liberalism, then, is neither an easy task nor a particularly desirable endeavour. However, any attempt at relating forms of restructuring to neoliberalism in order to see to what extent – if any – they can be seen as part of a larger neoliberalization process requires at least a tentative working definition.

There are, nevertheless, some general definitions of neoliberalism. Larner, for example, takes neoliberalism to signify 'new forms of political-economic governance premised on the extension of market relationships' (2000: 5). For Thrift, 'neoliberalism is a set of conventions or stories about the right ways to do things in order to succeed economically – as a firm, a country, a person or [. . .] a city' (1999: 276). This extension of market relationships is driven by the logic of competition and effectiveness. In its most commonly conceptualized form, neoliberalism indicates a shift from the welfare state to the logic of the market. Political governments focus more on economic efficiency and competitiveness, and less on, say, full employment and social welfare. Thus, the market, instead of the caring institutions of the (welfare) state, becomes the source of well-being. This conceptualization of neoliberalism as a 'policy framework' (Larner, 2000) implies a 'roll-back' of welfare state activities, and is usually concretized by policies of liberalization, privatization and deregulation.

But neoliberalism is more than a set of economic policies: it is a new form of 'political-economic governance' (Larner, 2000: 5); 'one particular way in which government is made possible' (Thrift, 1999: 276); a 'political rationality that both organizes these policies and reaches beyond the market' (Brown, 2003: 4). This is what Larner calls 'neoliberalism as governmen-tality' – a particular form of governance that constitutes individuals and institutions in compliance with the norms of the market, producing calcu-lating, individualized subjects responsible for their own well-being (or misery, for that matter), calculated, profitable economic and social policies aimed at encouraging – even requiring – competition, and institutions guided by the overriding objectives of competitiveness and efficiency. As Brown maintains,

> [N]eo-liberalism is not simply a set of economic policies; it is not only about facilitating free trade, maximizing corporate profits, and challenging wel-farism. Rather, neo-liberalism carries a social analysis which, when deployed as a form of governmentality, reaches from the soul of the citizen-subject to education policy to practices of empire. Neo-liberal rationality, while foregrounding the market, is not only or even primarily focused on the economy; rather it involves extending and disseminating market values to all institutions and social action, even as the market itself remains a distinctive player. (2003: 7)

Building on these observations, I take 'neoliberalism' as a political rational-ity, and define 'neoliberalization' as a particular form of restructuring guided by this political rationality premised on the extension of market relations that privilege competition, efficiency and economic success. Such a definition seems to me useful for three reasons. First, it emphasizes a process rather than a static condition. Second, it encourages an approach that does not reduce neoliberalism to the application of a set of economic policies. Third, it pays attention to practices that (re-)constitute spaces, states, subjects, individuals and institutions for the purposes of government in a particular way. It takes into consideration not only the various tech-niques and devices of government, but also the constitution of the objects and subjects to be governed. Thus, in my analysis of urban policy, I focus on the constitution of its spaces of intervention, their articulation with dif-ferent kinds of issues and problems, and changing forms of intervention and their legitimization. I see this policy as part of a larger restructuring of the French state, which involves new forms of statecraft and governmental practices, re-scaling of the state apparatus, and the production of new spaces – of regulation, intervention and containment (Larner, 2003; Peck, 2001, 2003).

As for the manifestations of neoliberalism at the urban level, three fea-tures are commonly cited in the literature as characteristic, and could be seen to reflect its economic, social and penal aspects. The first issue is the institutionalization of inter-urban and inter-regional competition through (neoliberal) urban policies based on the logic of the market. This is fostered through a variety of programmes that include place-marketing, enterprise zones, urban development corporations and public–private partnerships (Brenner and Theodore, 2002; Peck and Tickell, 2002; see also Jones and Ward, 2002 for a political economy, and Raco and Imrie, 2000 for a governmentality approach to British urban policy). The second issue relates to the socio-economic and socio-spatial manifestations of neoliberalism. Neoliberal strategies deployed in cities, it has been argued, sharpen socio-economic inequalities and displace certain groups whose presence is deemed undesirable (Hubbard, 2004; MacLeod, 2002). Urban neo-liberalism is deeply concerned with imposing a certain 'social landscape' on the city. The third issue follows from the second, and involves new and aggressive strategies of policing and surveillance aimed at particular groups and particular spaces (mostly city centres), criminalization of poverty and the increased use of the penal system (Peck, 2003; Wacquant, 2001).

But the question remains as to what it is that makes these orientations specifically 'neoliberal'. Economic strategies of the sort described above have already been effectively discussed under 'urban entrepreneurialism'. A concern with certain groups in the city and its public spaces is not distinctively neoliberal, nor is it a new occupation for urban governments. Moreover, one could be authoritarian without being neoliberal; authoritarian forms of urban governance do not necessarily follow from neoliberalism. In short, all of these commonly cited features of urban neoliberalism, when considered separately, could occur in political regimes that are not necessarily committed to such an ideology.

What makes urban policy orientations neoliberal, I believe, is the political rationality underlying them (premised on the extension of market values) *and* the active construction – not only laissez-faire – of the conditions in which such a political rationality can be disseminated, including markets, but also various institutional structures and practices that privilege competition, efficiency and economic success. Neoliberalism requires 'political intervention and orchestration' (Brown, 2003: 10) – through political institutions, law, policy, institutional practices and social norms – in order to encourage and facilitate competition, efficiency and rational economic behaviour on the part of the individual members and institutions of the society (Brown, 2003; see also Peck and Tickell, 2002: 395–6). Neoliberalism extends as a political rationality – and not merely as policy package – which requires political agendas that construct the conditions of its development, dissemination and eventually normalization. For example, Peck and Tickell (2002: 394) argue that the development and deployment of the political rationality of neoliberalism was crucial in reinforcing, extending and, more importantly perhaps, normalizing the consolidation of competitive urban regimes – either to 'win' in the inter-urban competition, or to secure a place in the global race. And when neoliberal rationality extends to the state, not only does the state respond to the needs of the market (through measures that range from monetary policy to immigration policy, from welfare programmes to the workings of the penal system), but also the criteria for its success and legitimacy get indexed to the market. The state is successful as long as it secures and promotes the market, the health and growth of which is now its responsibility and the basis of its legitimacy (Brown, 2003: 12–14).

The orientations referred to above may cohere around this political rationality rather nicely, consolidating a relatively coherent neoliberal regime – as, for example, in certain US cities – or, in other cases, while the political rationality of neoliberalism guides mainly economic policies, policies in other spheres follow other rationalities, producing hybrid and sometimes contradictory forms of neoliberalism – as, for example, in France. In order to understand urban policy in France, it is necessary to look at the transformations of the French state in a way that considers not only its contemporary restructuring along dominant political-economic rationalities, but, equally importantly, its established political traditions.

The Republican State and Its Contradictions

'Neoliberalism *à la française*', Jobert and Théret (1994: 80) maintain, is 'not a shameful neoliberalism. Its reconciliation with the Republic granted it a degree of authority and respectability.' They refer to this reconciliation as 'the republican consecration of neoliberalism'. Whether shameful or not, it is nevertheless possible to distinguish a French version of neoliberalism deeply influenced by the republican tradition emphasizing the social obligations of the 'republican state' towards its citizens. This is a conception of state that is highly endorsed by the public as well, which perhaps was best exemplified by the massive strikes in 1995 against the then Prime Minister Alain Juppé's attacks on the welfare state, and, more recently, by the immense opposition to a new labour law, the CPE (*Contrat Première Embauche*, First-Job Contract), which was seen to undermine job security for young people.[5]

The republican tradition is universalistic and assimilationist; it combines political membership (citizenship) with cultural membership (assimilation into 'French culture'), and emphasizes the role of the central state in actively promoting citizenship – mainly through an impartial technocratic bureaucracy – and securing 'solidarity', 'social bond' and 'social cohesion' (Feldblum, 1999; Silver, 1993; 1994). It, therefore, refers to one particular way of conceptualizing the relations between state and society, with particular emphasis on the duties and obligations of the central state vis-à-vis its citizens. It follows a social not an economic rationality (see Silver, 1994). Still influential in many ways, the French conception of the republican state highlights the state's obligations 'to guarantee citizens' social justice through the provision not just of traditional social services but also public infrastructural services' (Schmidt, 2002: 279). Such a conception not only prevented – to a certain extent, at least – social policy reforms along neoliberal lines, but also made it considerably difficult to legitimize such modifications since the political rationality of neoliberalism and that of the 'republican state' are logically contradictory. This contradiction has continued since the 1980s with belt-tightening economic policies *and* expansive social policies and services (Schmidt, 2002).

The Socialist government of the early 1980s (which initiated urban policy) had tried to implement the programme for which they were elected, which included increased state intervention and nationalization. Two years later, following the monetary crisis of 1983, already under pressure from financial markets and the EEC, the Socialist government started to move towards a neoliberal economic programme, adopting policies of budgetary austerity and privatization. This orientation continued with the successive governments of the left and the right through policies of financial market liberalization, privatization, business deregulation and labour market decentralization (Budgen, 2002; Jobert and Théret, 1994; Knapp and Wright, 2001; Levy, 1999; Schmidt, 2002). The application of such policies, however, carried the mark of France's (former) state capitalism. Despite market-oriented reforms in the last two decades, as Schmidt (2002, 2003) effectively argues, the French state has not moved towards market capitalism, but from 'state-led' capitalism to a sort of 'state-enhanced' capitalism, where the state still plays an active, if diminished, role. Furthermore, while implementing policies associated with neoliberalism, the French state has not completely retreated from its welfare functions. Instead, it has intensified its social interventions through public aid policies, minimum income programmes, government-sponsored work contracts and a universal health coverage plan (Levy, 2001; Wacquant, 2001). Increasingly since the 1990s, social policy has become a key issue for governments of both the left and the right (Levy, 2001).

The French state's 'neoliberalism', then, calls for reservations. Indeed, given the strong state tradition in France, neoliberalism as a political ideology has little or no place – at least to be voiced explicitly – even in the French right's political agenda. It should be remembered that before the dismantling of the *dirigiste* state in the 1980s, it was the French right that ran the *dirigiste* model for decades, and that the main party of the right (the Gaullist RPR) was founded on statist principles (Levy 2002).[6] Neoliberalism briefly entered the right's agenda as a political counter-ideology only after it lost power to the left in the early 1980s (Schmidt, 2002), and during Chirac's unsuccessful 1986–8 government. Even then, Chirac had promised not to touch the welfare state. And during his 1995 presidential campaign, Chirac (then head of the main party of the right, the RPR) had denounced the neoliberalism of his fellow party member Edouard Balladur (then Prime Minister), and organized his campaign around the theme of 'social fracture', which he would seek to heal through intensified state intervention. Before he reversed course a few months later with a stated aim to qualify for European Monetary Union by reducing the country's budget deficit, concrete measures indeed were taken, such as a 4% increase in the minimum wage, and financial subsidies to employers willing to hire unemployed youths (Levy, 2001).

The republican tradition, therefore, is not without influence on the contemporary restructurings of the French state, which differs remarkably from a US- or UK-style neoliberalization. Three issues, however, should be emphasized. The first relates to the emphasis on French cultural values, which is directly linked to political membership in the 'one and indivisible' republic. The republican tradition is 'far more intolerant of diversity in public life than American pluralism' (Silver, 1993: 346), and is prone to taking exclusive tones following from its obsession with culture. Exemplary in this sense is the message delivered by Charles Pasqua shortly before taking office as the Minister of the Interior in the centre-right government of 1993.[7] Pasqua stated that a 'multi-ethnic and multi-racial' society would be tolerable, but not a 'multi-cultural' one: 'If France does not suit them, all they have to do is go home and bugger off [foutre le camp]. [. . .] Those who want to live on the national territory must become French and assimilate our culture, we don't have to put up with the others' (Le Monde, 21–2 March 1993: 11). Such a tendency may easily lead to the demonization of 'other' cultures as 'threats' to national identity and cohesion, but also to the formulation of problems as following merely from cultural differences. This has partly been the case in the evolution of urban policy and the framing of banlieues, as we will see in the following chapters.

The second issue relates to the uses of republican rhetoric in political discourse ('social cohesion', 'solidarity', 'social bond'). Given the deep attachment to the republican tradition, such rhetoric is usually part of political agendas, in words if not always in deeds. It seems, indeed, 'the only way to win' even for right-wing governments, who may then shift to neoliberal agendas, as Chirac did in 1995 (Budgen, 2002: 32). But the reverse may also be true. The Socialist government of the early 1980s, for example, largely avoided republican rhetoric, and employed, instead, notions such as 'inequality' and 'social justice' – a strategy aimed at connecting issues to the structural dynamics of capitalism rather than to an organic conception of society (Silver, 1994). These notions, however, gradually disappeared from the political discourse after the 1983 turn of the government, and were replaced with the notion of 'solidarity', which, it was argued, implied a tacit acceptance of persistent inequalities (Jobert and Théret, 1994), and which could be seen 'as much as a way of buying off the most affected interests [by neoliberal economic policies] as the defence of traditional values' (Schmidt, 2002: 277).

The third issue follows from the previous two, and is particularly important given one of the lessons imparted by the French case: that although there are elements of convergence, the contemporary restructuring of the French state differs remarkably from a US- or UK-style neoliberalization partly because of the republican tradition emphasizing the active role of the state for the well-being of its citizens. Emphasizing the importance of established political traditions in the restructurings of states runs the risk of asserting a certain national exceptionalism, accounting for everything and nothing in particular. Therefore, I do not submit that the republican tradition has remained an historically continuous 'model' with coherent policy implications. Far from being an unchanging model, even the very meaning of 'republicanism' is constantly contested, and the term is employed by extreme right parties as well as anti-racist organizations calling for a 'true republicanism' (see Feldblum, 1994). As I mentioned above,

Socialists largely avoided the term in the early 1980s. Starting with the early 1990s, the republican conception acquired an exclusionary form with an emphasis on French identity and cultural values. Since then, the language of republicanism has continued under the right and the left, both emphasizing the authority of the state. In none of these periods, however, have the social services of the state been dismantled, although there have been attempts to do so. For these reasons, it is best to see republicanism as a relatively established political tradition that emphasizes the social duties and obligations of the state for the well-being of its citizens, not as a 'model' unanimously followed by succeeding governments.

Such an approach also requires an attentiveness to various 'republicanisms', and the reasons behind their (re-)appearance, such as the resurgence of republican nationalism in the 1990s. Many scholars working on contemporary France have observed a change in the 'attitude' of the republic towards its 'strangers'. This change consisted of a renewed enthusiasm for the republican tradition with nationalistic overtones – what I refer to as republican nationalism – and was most notably felt, it has been argued, in citizenship and immigration policies of the 1990s (Balibar, 2001; Blatt, 1997; Feldblum, 1999; Tévanian and Tissot, 1998). As we will see in the following chapters, this change also had a strong spatial dimension. French urban policy, just like citizenship and immigration policies, has been affected by this burgeoning republican nationalism since the 1990s with an emphasis on the 'values of the republic' and 'authority of the state' allegedly under threat from 'communitarian' groupings and the formation of ghettos in the *banlieues*. The French state's contemporary restructuring is also marked by the rise of republican nationalism, with a shift towards more authoritarian and exclusionary policies, and the consolidation of the 'republican penal state' mainly in and through the social housing neighbourhoods in *banlieues*.

The Republican Penal State and Urban Policy

That the contemporary restructuring of the French state involves a commitment to some form of neoliberalism has been observed by many scholars (Jobert and Théret, 1994; Levy, 2001, 2002; Schmidt, 2002; Wacquant, 2001). However, this restructuring also carries the signs of the strong state tradition in France, influenced by the idea of the 'republican state' and its social obligations towards its citizens. One of the lessons that the French case offers is that established political traditions affect forms of neoliberalization and state restructuring. I maintain, therefore, that such restructurings are best understood as articulations of 'the neoliberal project' with established political traditions, an articulation I try to capture, for the case of France, with the notion of the 'republican penal state'. While there are linkages and echoes suggesting that the French path is converging with a neoliberal one, there exist major tensions and contradictions deriving from inherited political traditions – of which an emphasis on 'the republic' and its duties is one.

The notion of the republican penal state is also aimed at taking into consideration variations in what Wacquant has identified as the 'European penal state', which follows the strong state tradition in Europe, and intensifies regulation through both social and penal policy-making. The 'left hand' of the state is still active, but it is increasingly accompanied by its 'right hand' through intensified use of the police, courts and prison system, and with a form of regulation following a 'panoptic logic' that involves the criminalization of the poor and the close surveillance of populations seen to be problematic (Wacquant, 2001).

The penal state, however, varies with different political traditions, deploying different containment strategies and legitimizing discourses. The 'new penal commonsense' (Peck, 2003) came to France with a republican twist, and shifted emphasis from prevention to repression through a legitimizing discourse organized around 'the republic' under threat by allegedly incompatible cultural differences and the formation of 'communities' unacceptable under the 'one and indivisible' republic. The republican penal state still has an active 'left hand', which, however, is increasingly accompanied by its 'right hand', concerning, in particular, the social housing neighbourhoods in *banlieues*. This is not to imply that the French state, after a decade of absence, is now back in deprived areas, as in the 'roll-out neoliberalism' of the US and UK cases (Peck and Tickell, 2002). The French state has been present in such areas through its urban policy (among others) for years – the 1980s included. The change between the 1980s and the 1990s, in this sense, was not so much the return of the absent state to spaces of poverty and mass unemployment as the remarkable change in the modes of intervention and discursive articulations of such spaces. It is this change that I chart in French urban policy.

Before moving on, let me summarize the main aspects of my approach. First, rather than taking the spaces of urban policy as given, I look at the ways in which they were constituted as objects of intervention through state's statements – policy discourses, spatial designations, mapping, categorisations, namings and statistics. Therefore, I see urban policy as an institutionalized practice of articulation that constitutes its spaces of intervention as part of the policy process, and consider the spaces of urban policy with the practices of their articulation.

Second, I look at the particular spatial order – 'the police' – consolidated by this articulatory practice and its political implications. I maintain, therefore, that urban policy is a place-making practice, but also that this

place-making is challenged by alternative voices formulated in the intervention areas of urban policy.

Third, I focus on the ways in which space is imagined and manipulated
in urban policy. I insist that different policy discourses and programmes
imagine space in varying ways, and that changing conceptualizations of
space and their policy and political implications should be taken into consideration in analysis of urban policy. As Estèbe (2001: 25) notes, the basis
of French urban policy is its definition of a 'geography of priority neighbourhoods': that is, a geography constituted by the designated areas, which
then becomes the basis of policy programmes and interventions. Based on
how such areas have been constituted, he identifies two such 'geographies':
a 'local' geography of priority neighbourhoods (in the 1980s) and a 'relative' geography of priority neighbourhoods (starting from the 1990s). I
follow Estèbe's analysis, but argue that a third, 'statist' geography was also
consolidated in the 1990s.[8] As we will see, the three themes around which
I organize chapters 3 to 5 in Part II – revolts, surveillance and repression
– correspond to these changing spatial conceptualizations – local, relative,
statist – as well as to changing forms of state intervention, from militant to
bureaucratic to authoritarian.

This brings me to the final aspect of my approach. I relate urban policy
to state restructuring, paying particular attention to dominant rationalities,
legitimizing discourses, established political traditions and major national
and international events. In other words, I situate my analysis of urban
policy in a broader political and economic context, showing how it feeds
down into urban policy. Such an approach provides insight not only into
French urban policy, but also into the contemporary restructuring of the
French state. It points to the role established political traditions play in
processes of neoliberalization, and offers insight into the making of the
geographies of the penal state through urban policy.

The main issues that French urban policy is concerned with, the very
issues that stimulated its conception and, later, institutionalization, have
not changed: incidents of social unrest – which have since intensified and
geographically extended – and the concentration of certain groups in certain
areas, a consequence mainly of the financial difficulties generated by the
massive job losses of the economic restructuring processes from the 1980s
onwards – which have since worsened. But the ways these issues have been
articulated as problems, representations of intervention areas, forms and
legitimizations of state intervention have changed remarkably over the
years, consolidating a 'police order' almost in a literal sense. This is not to
imply that urban policy-makers maliciously constituted these areas as zones
of containment and repression. But urban policy, as we will see, has been
tightly intertwined with issues of immigration and citizenship, anxieties
about French identity, 'values of the republic', and the authority of the

state. It thus evolved alongside the contemporary transformations of the French state on increasingly authoritarian and exclusionary lines. When urban policy was first conceived in the early 1980s *as a response* to revolts in the *banlieues*, it sought to create a political dynamic with such strong political ideals as 'democratization of the management of the city', 'appropriation of space by inhabitants' and 'the right to the city'. Two and half decades later, in the autumn of 2005, *banlieue* revolts were suppressed by unprecedented repressive measures with the declaration of a state of emergency. Well, how did we end up here?

Part II

The Police

3

The Right to the City? Revolts and the Initiation of Urban Policy

At the end of a chapter entitled 'Towards an urban strategy', included in his 1970 *La révolution urbaine*, Lefebvre identified three 'political strategies':

a) introducing the urban problematic into (French) political life, pushing it to the forefront;
b) elaborating a programme of which the first article will be *generalized autogestion* [. . .];[1]
c) introducing [. . .] the 'right to the city'. (Lefebvre, 1970: 199)

As if to follow Lefebvre, all these 'strategies' were in place in the urban policy programme initiated by the Socialist government shortly after coming into power in 1981 – not quite as pre-conceived political strategies but, rather, as an urgent response to incidents of unrest in the *banlieues* of several French cities. This policy was aimed mainly, though not exclusively, at the social housing neighbourhoods in *banlieues* through a spatial, rather than a sectoral, approach. Although originally conceived as experimental and not necessarily permanent, this policy, as we will see, would lay the basis of what would eventually become a permanent geography of intervention areas. It would, in other words, be the first step in the consolidation of 'the police', although its spatial approach was initially guided less by a governmental drive than by a desire to promote the right to the city in the designated areas.

The origins of urban policy go back to 1977 and the first programme of Housing and Social Life (*Habitat et Vie Sociale*, HVS hereafter). This programme was originally conceived to address the problems of large-scale social housing estates (*grands ensembles*) located in the peripheral areas of major cities. These estates were mostly built in the 1960s, and they started to show signs of degradation in the 1970s. In the 1960s, France was in the middle of its *trente glorieuses*: that is, the thirty years of economic growth from the end of the Second World War to the economic crisis of the 1970s, marked by rapid industrialization and urbanization. The *grands ensembles* were a quick, cheap and large-scale response to address the housing problem that had emerged in the post-war urbanization period. They were built mainly, as noted, in the peripheral areas of cities where land was available and cheap. They definitely contributed to the improvement of the lives of many families with their large surfaces, central heating, bathrooms and toilets, as Merlin (1998) explains in his book on the evolution of French *banlieues*.[2]

These housing estates initially did not accommodate deprived populations; they, rather, had inhabitants with stable incomes. By the end of the 1960s, each city, regardless of its size, had at least one neighbourhood composed of such housing estates (Jaillet, 2000), referred to as HLMs. The HLM, French social housing, was conceived for low-income families, although large numbers of middle-class families lived in such estates in the early post-war period. Most of the HLMs were constructed after the war in designated peripheral areas, the *Zones à Urbaniser par Priorité* (Priority Urbanization Areas, ZUPs hereafter). The ZUPs were created in 1958, and although the title was changed to *Zones d'Aménagement Concerté* (Concerted Planning Areas, ZACs hereafter) in 1967, the often negative image of the ZUP remained. As Hargreaves (1995: 70) suggests, the HLMs that characterize such areas may be seen as 'the French equivalent of British council housing and American housing projects'.

Although the development of such areas contributed to the eradication of shanty towns (*bidonvilles*), and improved the living conditions of many families, they were far from the city centres, and under-equipped. Most of them suffered from the lack of adequate public transportation, shops, social and cultural amenities, and from physical degradation due largely to the use of cheap construction materials and rapid construction techniques. The housing finance reform of 1977 opened the exits for middle-class populations with stable incomes, who were growing dissatisfied by the living conditions in these areas. The main objective of this reform was to facilitate owner-occupied housing. Those having the financial means took advantage of this reform and moved out from these areas. They were replaced by socio-economically disadvantageous populations; populations with unstable income and immigrants, who were hardly welcome by the property market in city centres. Moreover, rising unemployment exacerbated the problems in these areas. An inter-ministerial committee for housing and social life, the HVS, was established in 1977 in order to address these emerging problems. First experimented in 1972–3, the HVS programme was generalized in 1977 with the selection of fifty sites for housing renovation. The degradation of the *grands ensembles* was the main concern of HVS, which sought to address their problems through physical renovation. One critic argued that the major concern was to change the image of the selected areas with the hope of keeping the middle-class families there and attracting more middle-class populations (Aballéa, 2000). Another criticized the HVS programme for being 'too centralized, implying neither local elected representatives nor inhabitants, content often with re-doing a "new skin" to buildings without really improving the comfort of living conditions, and with extra "colouring" of the façades, which increase the growing or already affirmed stigma from which these neighbourhoods suffer' (Jaillet, 2000: 31). Jaillet's criticism of the HVS programme, however, seems rather harsh compared to the account provided by Merlin (1998: 134), who argues that although the main concern of the programme was the physical improvement of the housing stock, this went beyond a simple '"colouring" of the façades'. Furthermore, the economic crisis and the ensuing rise in unemployment had not hit the social housing neighbourhoods too severely at the time. Such issues, therefore, were not at the core of the programme, but they were not completely neglected either. The same seems to be the case for the participation of inhabitants, which was, indeed, one of the stated objectives – if unrealized – of the programme (DIV, 1999). Nevertheless, in the highly centralized administrative system, the state remained the dominant actor (Merlin, 1998).

The Hot Summer of 1981: How Novel is 'Violence'?

Urban policy was not simply an extension of the HVS programme. It was conceived following the incidents of unrest in the so-called 'hot summer' of 1981, a few months after the arrival of the Socialist government to power. Incidents mainly took place in the social housing neighbourhoods of the *banlieues* of Lyon, notably in Les Minguettes.

This is the routine account of the origins of urban policy and the introduction of *banlieues* in the political agenda. Such incidents, however, were not entirely novel; similar incidents had occurred in the *banlieues* before. In his chronology of urban policy, Daoud (1993: 136), for example, marks the year 1978 for the 'first incidents of violence in the *cités*'. Mucchielli (2001) mentions confrontations between the police and the youth in Vaulx-en-Velin as early as 1979. Bachmann and Le Guennec (1996: 329–30) show that Vénissieux, another *banlieue* of Lyon, was already a 'regular' of the local press with Vaulx-en-Velin and Villeurbanne (the so-called '3Vs') as the 'Lyon Bronx'. Similarly, Rey (1999) argues that incidents of unrest had occurred in *banlieues* more than once well before the hot summer of 1981.

These remarks suggest that an exploration of the specific context in which urban policy emerged is important in order to understand what was at stake. Before the arrival of Socialists in power, incidents of unrest in the *banlieues* had not escaped the attention of the right-wing government of President Valéry Giscard d'Estaing. Four years before the hot summer, in 1977, a commission chaired by Alain Peyrefitte, the then Minister of Justice, had published a report entitled *Réponses à la violence* (Responses to Violence, Peyrefitte Report hereafter). The conviction of the commission was that there had been a 'sudden rise' in violence in France, and that a 'feeling of insecurity' had appeared (1977, vol. I: 31 and 32). The Peyrefitte Report, while not neglecting other issues, linked violence to the city. Increasing violence and criminality, the report stated, was linked to a particular form of urbanization referred to as the 'disorder [*dérèglement*] of urbanization', which invoked the social housing estates built in the ZUPs. To account for the link between this form of urbanization and violence, three factors were identified. The first one of these was the 'cramming in' (*'entassement'*) of populations in high-rise buildings. The second factor was segregation, which was seen as a consequence of modern town planning principles based on the separation of functions. The third factor identified by the Peyrefitte Report was the anonymity of populations, which was seen to follow from the principles of modern architecture and its monotonous structures. This form of urbanization, according to the report, was closely linked to violence and criminality: 'Beyond six floors, criminality levels increase sharply. [. . .] [T]here exists a close link between certain forms of housing or urbanization (*grands ensembles*, high-rise blocks) and acts of violence against individuals' (1977, vol. I: 142–3).

The same year that the Peyrefitte Report was published, President Giscard d'Estaing charged his Minister of State for Immigrant Workers with a task to devise ways to reduce the immigrant population, targeting non-European, and especially North African, immigrants. Many young North Africans were expelled through the use of the discretionary powers of the Minister of the Interior, but Giscard d'Estaing was hoping for more. He made an attempt, shortly before the presidential elections of 1981, to repatriate 500,000 Algerians by force, but failed (Weil, 1991; Weil and Crowley, 1994). It was in this context that the Socialists came to power in May 1981. One of their promises was to suspend the expulsion of immigrants.

Brixton in France? The Haunting of the French Republic

Was France going to experience, in the same way as the United States or England, forsaken neighbourhoods and areas of uncontrolled social explosion?
Dubedout Report, 1983: 5

Our country has not experienced the explosions of violence that some working-class neighbourhoods abroad have been through in recent years.

Levy Report, 1988: 31

In France, large numbers of foreign workers were recruited to meet labour shortages during the 1960s and early 1970s. The oil shock and ensuing recession, however, altered the need to import labour. There was also a concern about the social and political implications of large-scale immigration. In 1974, organized recruitment of foreign workers was suspended. Family reunifications, however, kept the number of foreign populations growing, who were becoming more and more visible in everyday life (Brubaker, 1992).

In May 1981, Socialists came to power under the presidency of François Mitterrand with an ambitious programme that involved such politically fragile issues as the abolishment of capital punishment, suspension of the expulsion of immigrants, voting rights for immigrants in local elections, decentralization and *autogestion*. They were strategically able to keep all these promises as they came to power with a majority in the National Assembly and a strong base at the local level.

The principle of *autogestion* had served to bring together divided segments of the left, and was a strong rallying point while in opposition. However, once in power, the principle was abandoned due mainly to the conflicts between the Socialist Party (*Parti Socialiste*, PS hereafter), the French Communist Party (*Parti Communiste Français*, PCF hereafter) and major workers' unions. Furthermore, the negative effects of the economic crisis of the 1970s had rendered the working class more passive. There was a decline in labour militancy, and most workers, concerned with job security in a context of increasing unemployment, saw *autogestion* as 'utopian and even dangerous', and preferred more gradual reforms (Smith, 1987: 58).

Decentralization became the primary focus of the new government. Reforms were made with a stated aim to promote a 'new citizenship', as Prime Minister Pierre Mauroy put it, which involved a reorganization of the territorial governing structure of the state. The executive power of the prefect (the state's local representative) was transferred to elected assemblies in the departments and regions. Local governments were given tax-raising powers and more responsibilities. However, the communes were not addressed adequately in a context of deindustrialization, which hit some communes more severely than others. They were all treated on an equal basis, but the issue of resources available to them in addressing problems or performing required services was not carefully considered. This would have serious consequences since the negative effects of the economic crisis had started to be felt more severely, disproportionately affecting communes with mainly working-class populations. Many workers were directly affected by job losses in industry and manufacturing. They were also affected indirectly because the industrial *banlieues* where many workers lived started to lose their fiscal potential. In addition to this, urban renewal programmes and increasing rents obliged many workers in city centres to move away to peripheral areas, where increasing unemployment, decreasing fiscal resources and lack of facilities seemed to be concentrating (Le Galès and Mawson, 1994; Preteceille, 1988).

Another issue on the agenda of the new government was voting rights for immigrants in local elections. But this promise was not kept; the public, President Mitterrand argued, was not yet ready for that. The other promise concerning immigrants, however, was kept. The expulsion of immigrants was suspended, and measures were taken for the regularization of the status of 'illegal' immigrants. With a law passed in 1984, most foreign residents, regardless of their origins, were granted a renewable ten-year residency permit that would allow them and their families to settle in France (Weil and Crowley, 1994).[3]

The decision to suspend the expulsion of immigrants was immediately challenged by commentaries in right-wing papers. *Le Figaro*, for example, wrote: 'In neighbourhoods with a high density of North African population, the situation becomes explosive' (7 July 1981: 26). *Le Quotidien de Paris* was even more straightforward: 'Now that the expulsions are suppressed, they [young Arabs] are going to steal our cars and violate our daughters' (7 September 1981).[4] The attitude in some segments of the left did not diverge greatly from these positions. The French Communist Party (PCF) candidate Marchais opened his 1981 campaign by asserting that 'there was too strong a concentration of immigrants in the population' (*Le Matin*, 1 October 1981).[5] The PCF had already 'played the ethnic card' during the campaign for the 1981 presidential elections (Hargreaves, 1995: 182), and supported the actions of a communist mayor in Vitry who, on Christmas Eve 1980, had ordered the demolition of a hostel accommodating African workers.

The anti-immigrant comments in 1981 were probably conditioned, in part at least, by the incidents of that year's hot summer, notably in Les Minguettes neighbourhood of Vénissieux, one of the *banlieues* of Lyon. The left was in power for the first time in the Fifth Republic, after 23 years, with an agenda including politically contentious issues. In this context, such incidents had particular significance, especially for the opposition right, which had centred its critique on the 'soft' attitude of the new government towards immigration (Bachmann and Le Guennec, 1996; Mucchielli, 2001). Moreover, 'race riots' had occurred on the other side of the Channel, notably in Brixton,[6] and their ghost was haunting the French republic. Indeed, a clear message was delivered to the presidential candidates before the election, a few days after the start of Brixton revolts. In Lyon, a young person of Algerian origin, a priest and a pastor were on hunger strike in order to protest against the expulsion of immigrants by Giscard d'Estaing's government. A full-page open letter to the candidates was published in *Le Monde* on 17 April, asking the government to immediately end the expulsion of immigrant youth. Typed in bold, capital letters was 'Brixton in France?' The practice of the government was judged as a 'moral genocide' that had to be stopped. Otherwise, the letter stated, the government would have to 'assume the responsibility of triggering off explosions in France similar to those in Brixton'; it would only then become clear that 'the hunger strike of Lyon was the last chance of non-violence' (Figure 3.1).

Les Minguettes is a social housing neighbourhood in Vénissieux, in the Lyon agglomeration, department of Rhône, region of Rhône-Alpes (Figures 3.2 and 3.3). The negative effects of economic restructuring were severely felt in certain *banlieues* of the East Lyon Region, where Vénissieux is located. Between 1975 and 1982, seven firms were closed in this *banlieue* (Belbahri, 1984: 108, fn. 3). In this period, the number of unemployed people in Vénissieux more than doubled (from 1,253 to 3,287), which corresponded to a rise in unemployment rate from 3.8% in 1975 to 10.8% in 1982. The banlieue also lost 10,000 of its inhabitants in less than a decade (from 74,417 to 64,848).[7]

The incidents were not unique to Les Minguettes; there were incidents in other social housing neighbourhoods of the *banlieues* of Lyon as well. Throughout the summer, there were confrontations between the youth and the police in Vaulx-en-Velin, Vénissieux and Villeurbanne – the so-called '3Vs' (Figure 3.4). The events reached their peak in early July with 'spectacular' incidents such as '*rodéos*', in which cars were stolen and set on fire. Nevertheless, Les Minguettes became the symbol of the incidents of 1981. It became, as Belbahri (1984) argued, the symbol at once of the inadequacies of social housing estates *and* the position of immigration in France. Indeed, the incidents of 1981 helped bring to the surface an imminent debate on immigration in France, which was marked by the haunting ghost of the so-called 'Anglo-Saxon model'. Although '*race* riots' were unimaginable under the French Republic, unlike, say, Britain and the United States, the incidents nonetheless gave rise to fears.[8] As Dubet and Jazouli wrote: 'It was as if, suddenly, there had been a qualitative change in the perception of delinquency, French society had suddenly felt it was threatened by popular riots familiar to Americans and that have been known in England since 1974' (1984: 8; cited in Mucchielli, 2001: 106–7). French urban policy originated in this particular context, which led to the association of the *banlieues* with the issue of immigration. There were already concerns with immigration in the 1970s, framed generally with references

15 ème jour de la grève de la faim illimitée

de Hamid B., jeune « immigré »
Christian DELORME, prêtre
et Jean COSTIL, pasteur
à Lyon

pour l'arrêt immédiat des expulsions de jeunes «immigrés»

BRIXTON EN FRANCE?

LETTRE AUX CANDIDATS A LA PRESIDENCE DE LA REPUBLIQUE

Depuis le 2 avril, un jeune « immigré », Hamid B., un prêtre, Christian Delorme et un pasteur, Jean Costil, risquent leur vie dans une **grève de la faim illimitée** pour obtenir, à l'aube du nouveau septennat, la publication d'un texte juridique stipulant que « les enfants nés en France de parents d'origine étrangère, ou qui y ont passé l'essentiel de leur jeunesse, ne peuvent être **expulsés** du territoire français ».

Chaque année en effet — et celle-ci plus que toutes les autres — des milliers de fils et de filles de travailleurs immigrés, surtout maghrébins, qui ne sont eux-mêmes ni encore travailleurs ni, de fait, immigrés, sont bannis de ce pays où ils ont toutes leurs attaches et où ils tiennent à vivre, pour être rejetés vers des pays qui leur sont presqu'inconnus, même s'ils furent ceux de leurs parents.

En passant cette génération à la trappe, en la transformant en « génération boat-people », la France commet là un véritable génocide moral. Elle se rétrécit anachroniquement et devient une « terre d'apartheid » au moment même où l'évolution des sociétés développées va dans le sens de la multiracialité. Feindre de l'ignorer c'est prendre la responsabilité de déclencher ici des explosions semblables à celles de Brixton. On comprendra alors que la grève de la faim de Lyon était la dernière chance de la non-violence.

Nous attendons de vous que vous vous prononciez clairement et publiquement, avant l'échéance électorale, sur cette question brûlante qui engage le vrai choix de la société de demain.

> Les collectifs de soutien à la grève de la faim de Lyon
> Les signataires du « Manifeste contre la France de l'Apartheid »

DECLARATION DES EVEQUES D'ALGERIE

« Réunis avec le Cardinal Duval, les Evêques d'Algérie, profondément émus par le témoignage que portent l'abbé Christian DELORME, ainsi qu'un pasteur protestant et un jeune Algérien en faveur des jeunes « immigrés » menacés d'expulsion du territoire français, se déclarent solidaires de la cause qu'ils défendent parce qu'elle est celle de la justice et de la fraternité ».

> Alger, le 4 avril 1981.

Contacts à Lyon :   ACFAL, 68 rue Rachais, Tél (7) 872 98 58
Contacts à Paris : Journal « Sans Frontière », 33 bd St Martin, 75003 . Tél (1) 278 44 78 et 278 47 59
Soutien financier : MAN, ccp. 1 - 661 - 92 S Lyon (mention : solidarité jeunes « immigrés »)

Figure 3.1   Brixton in France? (*Source: Le Monde,* 17 April, 1981: 11)

Figure 3.2 Regions and major cities of mainland France (*Source*: Adapted from *Vaulx-en-Velin, ma ville . . .* (1996))

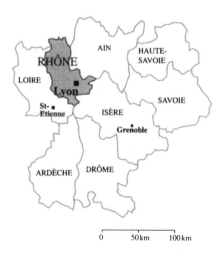

Figure 3.3 Departments and major cities of Rhône-Alpes (*Source*: Adapted from *Vaulx-en-Velin, ma ville . . .* (1996))

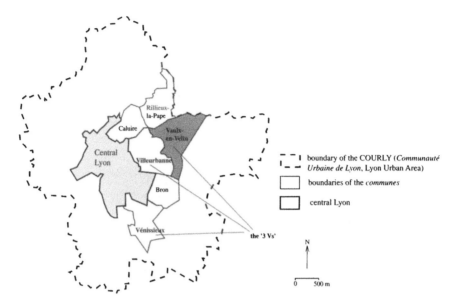

Figure 3.4   The '3Vs' of Lyon (*Source:* Adapted from *Vaulx-en-Velin, ma ville. . .* (1996))

to unemployment. The incidents of 1981 gave these concerns a spatial form and put 'the *banlieue*' on the political agenda.

Before moving on to urban policy, it might be helpful to remember some of the points regarding the context in which it was conceived. The incidents of Les Minguettes were not novel, nor were they generated out of thin air. However, their form was novel and more 'spectacular' compared to previous incidents. Sylvie Harburger was a researcher in the Ministry of Public Works at the time, and she joined the urban policy commission created after the incidents. Here is how she remembers the impact of the 'hot summer' of 1981:

> Well, you know, they were received in a way . . . even worse than the riots in November last year [2005] [. . .] because it was the first time, you know. It was the first time and nobody had . . . no one in the general public had anticipated them . . . and it was a conflict, which incidentally is . . . strange, looking back. The conflict in 1981 was between the youth and the police. So in a way it's the very same conflict that hasn't been solved to this day. [. . .] You know, it was the first demonstration that got wide media coverage. I guess it wasn't any worse than what went before, but cars were burned down. (Interview, Sylvie Harburger)

Although the incidents came as a big surprise to the general public, the problems and tensions in *banlieues* were not totally unknown to the state and the media. Various state institutions, such as the *Commissariat Général du Plan* (the official organization responsible for economic planning) and the Ministries of Public Works, National Education and Justice, were already aware of the problems:

> While the urban crisis in a way 'surprised' local and national politicians during the summer of 1981, it was no novelty for civil servants working in services surveying city evolution. They had not foreseen the violence of the conflict (between the young people and the police), but they had a foreboding of the turning point, and they had identified it already in the late 1970s. (Harburger, 1994: 387)

Following from these remarks, the initiation of urban policy was not so much a straightforward response to a newly discovered problem. The 'problem', first, had to be articulated as a problem, and solutions proposed accordingly. The response of the new government carried immense strategic and symbolic significance. Strategically, the new government could not have afforded the recurrence of such incidents, especially when faced with the right's critique of its 'soft' attitude towards immigration. Symbolically too, because the Socialist Party's success was based largely on the emergence and activism of a new generation of militants and elites, mostly committed to urban struggles. Indeed, the initial impetus for the decentralization reforms 'had much to do with the pressure of the social movements that produced the 1981 victory' (Preteceille, 1988: 415). This meant that the Socialist government was very sensitive to social movements – indeed it had to be in order to maintain its locally based power (Harburger, 1994; Preteceille, 1988). Nevertheless, it was a bit perplexed since Les Minguettes conformed neither to traditional forms of urban struggles, nor to the formal associational forms, both of which the left was familiar with. There was, for example, no explicit claim made by the revolting youth (Donzelot and Estèbe, 1993; Jazouli, 1992).

There is surprisingly little account of what happened in Les Minguettes in the summer of 1981. There was apparently no specific incident that triggered the incidents, which consisted of *rodéos* and confrontations between the police and the youth (see, for example, the media review report prepared by the CNV, 1991b). A more detailed account is provided by Jazouli (1992), who argued that although no explicit claims were made by the young people, the incidents were provoked by a 'feeling of exclusion', generated by economic difficulties, failure at school and tensions with the police. Two interviews cited by Jazouli may support this interpretation:

> It's when the cops started their provocations that the young people became aggressive, because they didn't understand the aggressiveness of the cops

toward them. *Rodéos* are an answer to everything they have endured, them and their parents. They got hit, and didn't understand; the anger they had within, they took out on the cars. First it was big cars stolen in the city centre, everything they are deprived of; and because when you steal a car, who comes to get it? The cops! It was a little like a cat-and-mouse game, the objective was to attract the cops who wouldn't let them live in peace, to fight them in the evening, face to face. (Djamel, Les Minguettes; cited in Jazouli, 1992: 21–2)

For us, *rodéos* were a way of expressing how fed up we were, nobody could have foretold it would go so far; media commented them both negatively and positively, but no one could claim not to know what we were experiencing after that. It's only afterwards that that leftists and intellectuals turned up to try and explain to the guys the true meaning, as they said, of the *rodéos*. For us, it was bullshit, and, anyway we didn't get it. They weren't the ones who were deep in shit, their friends weren't in jail. They got it quite quickly, and some of them even got slapped in the face, it may have been stupid, but you have to understand that we were fed up with everybody. (Farid, Les Minguettes; cited in Jazouli, 1992: 24)

By the end of the summer of 1981, some 250 cars had been stolen and set on fire in the peripheral social housing neighbourhoods of Lyon, Marseilles, Roubaix, Nancy and Paris (Jazouli, 1992). These manifestations were taken seriously by the Socialist government, and an urban policy programme was initiated, which was unprecedented in many ways. In the words of Dominique Figeat, who participated in this programme:

[T]he left-wing government was very uneasy about this [i.e. the incidents of 1981]. So I think it had a crucial effect on the will of the Prime Minister at the time, Pierre Mauroy, to engage into, in actions of such importance, in the sense that, faced with those riots there could have been a purely securitarian or police-oriented response, but there was also a response that was urban, social, more political, more general, which was clearly what Pierre Mauroy wished for. (Interview, Dominique Figeat)

The 'Founding Texts' of Urban Policy

Following the incidents of 1981, a National Commission for the Social Development of Neighbourhoods (CNDSQ)[9] was created by Prime Minister Mauroy in the same year. In his letter dated 25 November 1981 addressed to Hubert Dubedout, mayor of Grenoble, Mauroy charged the commission with addressing 'the problem of social housing neighbourhoods', and attached it directly to the office of the Prime Minister. Three reports were prepared in the following two years, which are usually referred to as the 'founding texts' of urban policy. Before moving on to these reports, however, let us try to get a sense of what the climate was like among urban policy-makers at the time. They were, on the one hand, faced with a politically charged and urgent issue. On the other, they were given an opportunity to tackle the urban question with the full support of the government in, perhaps, unprecedented ways. How did the commission interpret its mission? What ideals did its members seek to promote?

Sylvie Harburger, who had joined the CNDSQ right after its creation, provides an account. There existed, she states, a peculiar 'state of mind', 'a real "burst" of enthusiasm' (1994: 385). Another member of the commission, Pierre Saragoussi, defines it as 'euphoric' (interview with the author). At the time, the CNDSQ represented a real innovation, and there was great hope and certitude that doing things 'differently' would help to ameliorate the situation (Harburger, 1994: 385). The nomination of Dubedout as the president of the commission was not merely a highly welcome decision; it also had immense symbolic value. It was highly welcome because Dubedout was nationally known (which was pretty uncommon for a mayor at the time), not for his political influence at the national level, but for his local practices as the mayor of Grenoble, which was then regarded as 'a laboratory of municipal management' (Harburger, 1994: 386). Dubedout was one of the founders of Municipal Action Groups (*Groupes d'Action Municipale*, GAMs) in the 1970s, which sought to promote *autogestion* at the urban level. He was also an advocate of voting rights for immigrants in local elections. Here is how Bachmann and Le Guennec describe Dubedout and the team he brought together: 'The small team gathered around Dubedout, as Dubedout himself, is heir to the urban movements of the 1970s. Therefore, it attaches extreme importance to local democracy, to listening to inhabitants, even if they do not vote, even if they are not organized in officially recognized associations' (1996: 378). Dominique Figeat was part of the team, and here is how he describes Dubedout and his colleagues:

First of all, the image of a mayor very involved in running his city, and the image of a man with a background in associative movements. And he himself wanted the measures taken for the urban transformation of the neighbour-hoods, to involve the associative movements, the social movements, and for it to take place in very close partnership between the public authorities, the state, the local authorities and the associative movements. And at the time, the issue of the implication of inhabitants in the projects was an issue that had come up. (Interview, Dominique Figeat)

The nomination of Dubedout was also symbolically important given the French state's established practice of top-down, technocratic interventions. Dubedout was not a technocrat but a locally elected official. Perhaps this might not sound a breakthrough now, but at the time, such a nomination was seen as rather astonishing: 'It was a revolution, not simply a novelty' (interview, Sylvie Harburger).[10] Dubedout's appointment was consistent with the new government's attempt to alter overly centralist state practices, exemplified by the decentralization reforms. 'We had the feeling we were inventing a new state profession, closer to and more respectful of other partners, notably locally elected officials' (Harburger, 1994: 386).[11]

Therefore, the particular 'state of the mind' of the early 1980s among the members of the commission, as Figeat, Harburger and Saragoussi described it, was characterized by three major orientations and sensibilities. First, there was an orientation towards people working in the field ('acteurs de terrain'), called the 'foot soldiers of democracy' by the members of the commission. Second, there was a sensitivity towards urban social movements, which probably was influential in the creation of the CNDSQ in the first place. Finally, priority was given to local officials rather than to technocrats; the commission members were much more sensitive to local specificities and experiences compared to the technocratic practices that had characterized earlier periods.

The Dubedout Report, *Ensemble, refaire la ville* (Together, Remaking the City), was published in 1983, proposing a new approach: Social Development of Neighbourhoods (*Développement Social des Quartiers*, DSQ hereafter).[12] This was accompanied by two other reports: the Schwartz Report, on the economic 'insertion' of the youth, and the Bonnemaison Report, on delinquency, published in 1981 and 1982, respectively. These three reports, the 'founding texts' of urban policy, addressed the aggravation of problems in the social housing neighbourhoods located in the peripheral areas of cities.

The problem identified by the Schwartz Report was that young people were the most affected group of the population by increasing unemployment. They had difficulty in finding jobs, and were eventually 'marginalized'. The objective of the report was to conceive ways for the better incorporation of young people between 16 and 21 into the job market. Spatially, the focus was on the *grands ensembles* and the social housing neighbourhoods with problems, described as higher levels of unemployment compared to the central city, low revenues of inhabitants, low levels of qualification, poverty (economic, social and cultural), over-density of population, inadequate public transportation, a high proportion of young people and a concentration of immigrant families.

With the Schwartz Report, the issue of second-generation immigrants started taking its place in urban policy-making. The report connected the problems of immigrant youth and social housing neighbourhoods in ZUPs, the negative image of which contributed further to their 'marginalization'. It was proposed that the central city and the periphery be considered together: 'In order not to enhance the ghetto impression, it is necessary to "qualify" the *grands ensembles* as neighbourhoods of the city' (1981: 137). The report called for a comprehensive approach that would include rehabilitation, employment and cultural activities. It also introduced a new term by proposing 'a *positive discrimination* in favour of young people' (1981: 30; emphasis added).

The Bonnemaison Report on delinquency was also concerned with the youth and immigrants, stating that immigrants, together with nomads and 'marginals', were the 'particularly disadvantaged' groups of the population. However, the report stated, '[t]he right to difference ha[d] to be not only recognized but respected' (1982: 48). Bonnemaison was the chair of the Commission of Mayors on Security, established in 1982. The report was aimed at defining the principles of a prevention policy, which emphasized proximity and collaborative action, with a concern to avoid an 'anti-youth' approach. Like the Schwartz Report, the Bonnemaison Report also addressed the issue of immigrants and social housing neighbourhoods. There was, the report argued, a lack of an active social housing programme, which resulted in the concentration of immigrants in certain neighbourhoods. The report proposed giving local officials power over housing policy, a fair distribution of social housing between communes, and 'furthering the harmonious distribution of immigrants in social housing programmes' (1982: 210).[13]

The report also provided an evaluation of 'Operation Summer 1982', which was conceived to 'prevent the recurrence of the events of the summer of 1981 like those in the *banlieue* of Lyon' (1982: 167). The 'operation' consisted of cultural and sports activities for young people, visits to other regions and summer camps, but also of a reinforced police presence. Another 'hot summer' would have serious and hardly desirable political consequences for the government. The summer of 1981 had made such an impression that in the following years, the approach of summer had become a major preoccupation among government and local officials. In the words of Gilbert Carrère, the regional prefect of Rhône-Alpes region at the time: 'Let us specify once more that the explosion of the summer of 1981 at Les Minguettes struck minds so strongly that, for years later, the approach of summer still obsesses administrations and locally elected officials, and has them preparing, more widely each year, summer operations which gather hundreds of participants for days or weeks' (1994: 390).

These reports show that from the start, urban policy was concerned with issues of immigration and social housing neighbourhoods. The third 'founding text' of urban policy, *Ensemble, refaire la ville*, which, as noted, was prepared by Dubedout and his team and published in 1983, clarified the orientations of the National Commission for the Social Development of Neighbourhoods (CNDSQ). The problems to be addressed by the commission were defined as insecurity, degradation of the housing stock, deterioration of social relations in the city, unemployment, educational problems and the difficulties facing minorities in terms of social and cultural 'insertion'. Faced with these problems, the objective was to start economic and social development plans in selected neighbourhoods, and to devise a decentralized and comprehensive policy. The field of action was defined mainly as the *grands ensembles* of social housing neighbourhoods, although some older neighbourhoods in city centres would also be considered. Initially, 16 neighbourhoods were selected (including Les Minguettes) for the Social Development of Neighbourhoods programme (DSQ).

The Dubedout Report opened by referring to the incidents of 1981 in Les Minguettes, and the others that followed in the northern neighbourhoods of Marseille, social housing neighbourhoods of Roubaix and Nancy, and the *cité* of 4000 in La Courneuve in the Paris region. The report was highly critical of the 'ghetto images' diffused by the media covering the incidents, and the 'designation of young immigrants as scapegoats' (1983: 5). The representation of social housing neighbourhoods as places of criminality was criticized, and it was suggested that measures be taken in this respect. The report stated that the young people who came to the centre of attention following the incidents (i.e. second-generation immigrants) were almost the same age as these neighbourhoods; they had lived there through their evolution, and experienced, therefore, 'the inadequacy of schools, teachers, facilities, leisure activities' (1983: 61). These were rather acute problems when social housing neighbourhoods in the form of *grands ensembles* were first constructed in the 1960s and the early 1970s. Young people living in these neighbourhoods not only suffered from the inadequacies of earlier years, but also from contemporary problems such as lack of training, exclusion from the job market, and the hostility of adults. The economic crisis of the early 1970s had exacerbated the problems: 'The development of intolerance, rejection, and racism profoundly affects a society already undermined by the [economic] crisis. The increase in an unfounded sense of insecurity is a disquieting symptom of this. If the current policy fails to prove effective, most citizens will call for authoritarian solutions' (1983: 98).

The conviction was that the crisis was not merely economic. It was also important to recognize the presence of minorities living in degraded social housing neighbourhoods, and the difficulties they encountered in economic, social, political and cultural terms. These difficulties, the Dubedout Report insisted, should be seriously taken into consideration, and the manifestations of young people be considered carefully, rather than interpreted merely as acts of delinquency.

While recognizing the presence and problems of immigrants in social housing neighbourhoods, the Dubedout Report above all emphasized the 'working-class' and 'popular' nature of social housing neighbourhoods, and insisted that they not be characterized in solely negative terms:

> The institutions must accept the reality of these neighbourhoods as popular neighbourhoods. Places of conflict, but also of solidarity, places of material poverty, but also of proliferation of popular cultures, places of difference, but also of blending, places of reject, but also of social insertion. (1983: 57)

By emphasizing the popular and working-class nature of social housing neighbourhoods, the Dubedout Report not only highlighted the relationship between the problems in such neighbourhoods and larger economic dynamics, but connected them to a certain political culture as well.[14] These working-class neighbourhoods were severely affected by economic restructuring, and immigrants and their families, along with unskilled workers and female-headed households, constituted the most precarious groups living in such areas.

The propositions of the Dubedout Report followed from an analysis of the previous HVS programme, organized around four critiques (1983: 9–10). First, the HVS programme focused, it was argued, too much on the physical improvement of housing while neglecting social and economic aspects of the problems of social housing neighbourhoods, such as education and employment. Second, inhabitants were not effectively included in decision-making processes, which created a suspicion on their part towards the projects being implemented. Third, locally elected officials, equally, were excluded from decision-making processes. When the *grands ensembles* were first conceived, the report argued, they were often imposed upon the communes by the planners of the central state, and the possibilities for the intervention of locally elected officials remained very limited. Finally, the report criticized the administrative procedure, arguing that it was 'heavy and too rigid'. In addition to these critiques of the HVS programme, the Dubedout Report maintained that partial and sector-based policies, while important, failed to take into consideration the social dynamics of the neighbourhoods – hence the commission's insistence on a spatial approach. Based on these critiques, three major orientations were defined for the DSQ programme.

First, it was necessary to act on the deprived neighbourhoods and the causes of their degradation in order to tackle problems generated by the concentration of disadvantaged groups in such areas. The deterioration of the social housing stock was only one side of a complex problem, which had to be addressed in ways that went beyond an exclusive focus on physical improvement. Although physical improvement was important (and the

HVS programme was given credit for that), issues such as unemployment, poverty and school failure had to be taken into consideration.

The second orientation concerned the participation of inhabitants. 'No progress can be accomplished', the report insisted, 'without a real appropriation by inhabitants of their environment' (1983: 75). The HVS programme, again, was given credit for attempting to achieve participation of inhabitants, but was criticized for focusing exclusively on formal associational forms, neglecting the 'diversity of forms of participation'. 'There is not only one but several levels of intervention of inhabitants in democracy at the scale of the neighbourhood' (1983: 38). The quest was for 'democratization of the management of the city' ('*démocratiser la gestion de la ville*'):

> In this quest, there are no privileged professions, no monopoly of knowledge, no hierarchical level, no command centres. Meeting this challenge requires a mobilization of all those who want to give access to *the right to the city* to all those from whom it is withheld. This dynamism must be part of a new political and technical organization of all authorities who manage these issues. For the moment, many have seen in the crisis the economic aspects, the restructuring of industrial sectors, the closing down of factories, the laying off of workers. The time has come for all those who, in one way or another, manage the city to understand that the crisis is as cultural, social, urban as it is economic. (1983: 29; emphasis added).[15]

The third orientation of the DSQ programme concerned the role of local collectivities. The proposition was to reinforce their role, giving them more power in larger decision-making processes, while increasing their responsibilities.

Although these three 'founding texts' of urban policy were conceived to address different concerns, they nevertheless shared two common features. First, all expressed concerns about the concentration of disadvantaged groups and immigrants in certain social housing neighbourhoods in the peripheral areas of cities. The often negatively connoted term '*banlieue*', however, was never used to refer to such areas. Second, these reports, while advocating a spatial approach that would target selected areas, pointed to the limits of focusing exclusively on selected neighbourhoods. The Schwartz Report, for example, insisted that the central city and its peripheral areas be considered together, not as separate. Similarly, the Bonnemaison Report emphasized the importance of conceiving prevention policies on a scale larger than individual neighbourhoods. The Dubedout Report went even further: 'The opening up of social housing neighbourhoods implies an ideological breakthrough: their inclusion in debates over the development of cities' (1983: 76).

The DSQ programme started with these orientations.[16] The first 16 neighbourhoods were selected in 1982, and seven more were added the following year (Figure 3.5). There was not a standard, 'objective' or formulized selection process; selections were made by the CNDSQ working in consultation with mayors. The members of the commission already knew the social housing neighbourhoods well. The selection was based on their knowledge of these neighbourhoods, taking into consideration their specificities: 'So, each time there was something specific. [. . .] So, in territorialization [of urban policy], there was also this notion of specificity, and of balancing things out (interview, Pierre Saragoussi).[17]

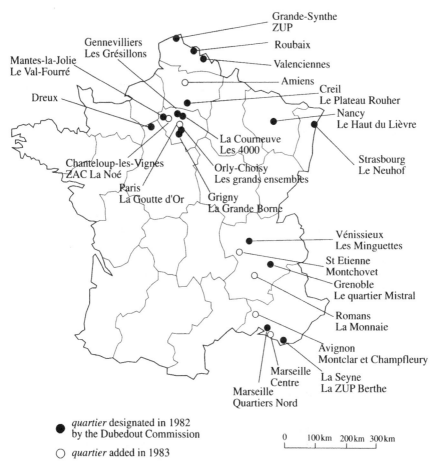

Figure 3.5   The first urban policy neighbourhoods (1982–3) (*Source: Délégation Interministérielle à la Ville* (DIV))

Another initiative of the same period was the creation of *Banlieues 89* in 1983.[18] While the DSQ programme criticized the exclusive focus on physical renovation, President Mitterrand, paradoxically, initiated this new programme under the direction of two architects, Roland Castro and Michel Cantal-Dupart. The major concern of the programme was physical rehabilitation ('the democracy of beauty' was one of their slogans). Some 600 architects were mobilized to work on more than 200 projects. The programme had an unprecedented media coverage, so much so that the architecture journal *Architectes architecture* devoted an article to the programme entitled '*Banlieues 89*: Trap for the media or policy?' (Loubière, 1984). Merlin (1998: 136–7) argued that the programme had been an occasion for a retaking of power by architects, with not too much worrying about the needs and expectations of inhabitants. It was visually more demonstrative of the government's effort, but, in a sense, it was 'an anti-DSQ'. As two of the members of the DSQ commission put it:

[W]e talked about neighbourhoods. So, a neighbourhood implies there's a city! You see? And so . . . it was necessary to connect . . . by talking of *banlieue*, they were closing in. That wasn't Castro's will at all. [. . .] But I think they put a curse by using that, that term, *banlieue*, because I believe . . . well now I'm jumping forward, I think the future of these neighbourhoods cannot be separated from the, well from the rest of city really. (Interview, Pierre Saragoussi)

Banlieues 89 was an elite of architects and urbanists, what you could call a technocracy of architectural and urbanistic action, projecting their representations of, well, both the existing populations and the existing neighbourhoods, on the formulation of policies. That was the absolute reverse of what Dubedout had tried to found. (Interview, Dominique Figeat)

Shortly after the 'hot summer' of 1981, the DSQ programme was in place to tackle the problems of peripheral social housing neighbourhoods with such political ideals as the democratization of the management of the city, appropriation of space by inhabitants, and the right to the city. The political context in which the problems of such neighbourhoods was addressed, however, would change significantly.

The 'Anti-immigrant Vote'

Jean-Marie Le Pen, president of the French extreme right party *Front National* (National Front, FN hereafter), obtained 11.3% of the votes in the 20th district of Paris in the municipal elections of 6 March 1983. While Le Pen became a member of the municipal council of Paris, 30 miles to the west of the city, in Dreux, another FN candidate, Jean-Pierre Stirbois, obtained 16.5% of the votes, which was unprecedented for the extreme right.[19] This was the first electoral 'victory' of the FN, to be followed a year later during the European elections with a score of 11% for all of France. The FN campaign was practically based on the argument that there were 'too many' immigrants, an argument that was also linked to the theme of insecurity. The opposition right's discourse did not diverge greatly from the extreme right's,[20] and the FN actually made local alliances with the parties of the right – Jacques Chirac's *Rassemblement pour la République* (Rally for the Republic, RPR hereafter) and the Giscardian confederation *Union pour la Démocratie Française* (French Democratic Union, UDF hereafter). These alliances helped the FN, as well as the RPR and the UDF, to gain local power, notably in the regional councils following the parliamentary elections three years later.

The increased popularity of the FN had much to do with the severe effects of the economic crisis and restructuring – the number of unemployed people had doubled in a period of five years, rising from 1 million in 1977 to 2 million in 1982. As Hargreaves (1995: 184) argues, although economic insecurity is but one issue of concern among others, such as the alleged erosion of French identity, 'it would be a mistake to underrate the significance of economic concerns among FN voters'. The 1983 elections took place in a context of growing disquiet about unemployment, with references to immigrants as 'surplus' or 'excess' when there were fewer jobs available. The 'anti-immigrant vote' (*Le Monde*, 13–14 March 1983: 9) was a product of this unease coupled with anti-immigrant discourses. However, neither anti-immigrant discourses nor rising unemployment was an entirely novel issue in 1983. Similar discourses were also mobilized in the 1970s, and the number of unemployed people had doubled also from 1973 to 1977. What happened in the 1983 elections, for some reason, had not happened in the 1970s in response to the anti-immigrant discourses of the FN and the French Communist Party. As Feldblum wrote:

> Le Pen and the National Front did not suddenly emerge in the eighties, rather they became electorally visible during this period. Their same discourse and platform failed in the seventies. Anti-immigrant rhetoric was also not greatly successful for the French Communist party (PCF) in the seventies, when they attempted to propel immigration themes into electoral issues. [. . .] Yet the proportion of immigrants in France did not change dramatically from one decade to another. (1999: 37)

There was not a spectacular increase in the proportion of immigrants in the total population from the 1970s to the 1980s (6.5% in 1975 compared to 6.8% in 1982). However, immigrants had become more 'visible' in everyday life starting from the end of the 1970s (Brubaker, 1992; Hargreaves, 1995). Until the early 1970s, men came to work mainly without their families, and they usually lived in hostels, and remained apart from the French. With subsequent family reunifications, immigrants became more visible in everyday life, in the housing market and in schools.

> In this way, immigrant groups which had seldom been encountered outside the workplace became visible on a daily basis in a growing number of neighbourhoods. Their increased visibility would not, of course, have been so marked had it not been for one other crucial point: far more than earlier generations of immigrants, those originating in Third World countries were instantly recognizable because of their skin colour and other somatic features. (Hargreaves, 1995: 19)

More importantly perhaps, immigrants had become more vocal. The new generation of immigrants were different from their parents in that they managed to show a political capacity and formulate political claims. In the words of Pierre-Didier Tchétché-Apéa, an activist in Vaulx-en-Velin (see Chapter 6):

> [O]ur claims or our demands are also different from our parents', in all respects. In behaviour, in visibility, in expression, in the claims, and the demands we have as citizens and so on, it's not the same at all. [. . .] So if, if what we are being reproached with, is our visibility, it's because we are more and more visible. That's what we're being reproached with! And our parents, they'd keep a low profile [*ils rasaient les murs*], our parents, they'd say 'yes Sir', and there you are. And we, well, we don't, OK? We walk head on, we don't hug the walls [*on rase pas les murs*]. (Interview, Pierre-Didier Tchétché-Apéa)

The early 1980s was also a period of rising immigrant activism.[21] Foreigners obtained the right of forming associations with a 1981 reform of the Mitterrand government, after which the number of associations 'mushroomed' (Hargreaves, 1995: 89). The incidents of Les Minguettes had sparked new forces of activism among second-generation immigrants, in particular those from North Africa. In the winter of 1983, a highly publicized event took place. A march started from Lyon (originating in Les Minguettes) and ended in Paris. This 'March for Equality and Against Racism' consisted largely of second-generation North African Immigrants ('*beurs*'). They asked not only for equality, but also that justice be done against racist crimes (see *Le Monde*, 4–5 December 1983: 1 and 11). The summer of 1983 was marked by racist killings targeting mainly young people of North African origin, claiming about forty lives (Dubet, 1987). President Mitterrand received a group of organizers, and announced the generalization of residency permits of ten years. He also stated that voting rights for immigrants remained still a 'preoccupation'. However, he maintained, public opinion did not seem to be ready to accept such a measure (*Le Monde*, 6 December 1983: 12).

The municipal elections of 1983 and their aftermath seem relevant to urban policy for three reasons. First, urban policy, as we have seen, is tightly linked to the issue of immigration. With the anti-immigrant vote, the 'legitimacy of the immigrants' presence in France, and their relation with French society became appealing targets' (Feldblum, 1999: 42). A nationalist-populist discourse appeared on the right and sections of the left, presenting immigrants as a 'threat' to the integrity of French nation and identity.

Second, the left's power at the local level diminished considerably. The left had gained strong local powers in the municipal elections of 1977, especially in larger communes with more than 30,000 inhabitants. In the 1983 elections, half of these larger municipalities were lost. Thus the decentralization reforms of the left benefited the right by giving it more power at the local level, where the left was starting to lose its base. Therefore, when the right returned to power after the 1986 legislative elections, it left decentralization largely untouched, and even asked for further powers at the local level (Mazey, 1993; Preteceille, 1988).

Finally, a right-wing candidate, Alain Carignon (RPR), was elected mayor in Grenoble, replacing Hubert Dubedout. This electoral defeat had immense symbolic power, and Dubedout immediately resigned from his position as the president of the National Commission for the Social Development of Neighbourhoods (CNDSQ). Rodolphe Pesce, mayor of Valence, was assigned as the new president of the commission. Dubedout left Grenoble, and settled in Paris with bitter memories of the 1983 municipal elections. During the campaign, he was attacked on his alleged Kabyle origins, and was accused of having changed his name to a more French-sounding one. Here is how he described the campaign:

> The campaign took place in the context of a considerable anti-immigrant racism. They said my mother was Kabyle. There is nothing you can do against a rumour. The situation reminds me of the rise of fascism in the 1930s: unemployment, a hard-line right, the poor white settler [*le petit blanc*]. [. . .] For instance, I received letters demanding that I prove that my mother was not Kabyle. In 1940, some French people were cowardly enough to declare publicly that they were not Jewish. Some people in the Socialist Party would have wanted me to make a public statement. I would rather lose an election than to give way to cowardice. (Cited in Parent and Schwartzbrod, 1995: 50)

The so-called 'anti-immigrant vote' of 1983 and its aftermath was a turning point for urban policy. In 1984, Prime Minister Pierre Mauroy was replaced by Laurent Fabius, who, as Dominique Figeat explained, did not support the DSQ programme in the way Mauroy did. President Mitterrand, similarly, was more supportive of image-oriented actions such as the *Banlieues 89*. Although the initial ideals of DSQ were not completely abandoned, 'there was a tendency [. . .], in a way, for the traditional state, already, to return to its old habits' (interview, Dominiqe Figeat).[22] This tendency would become more marked a few years later, under the Rocard government.

Consolidation of Urban Policy

By the time of Dubedout's departure, the number of neighbourhoods included in the DSQ programme had been raised from an initial 16 to 23 (see Figure 3.5 above), covering 300,000 inhabitants and 90,000 housing units. The selection process was still locally based; municipalities made proposals, and the final decision about which neighbourhoods would be included was made by the commission (CNDSQ) after taking the opinion of the prefect of the department in which the neighbourhoods were located. The geography of intervention areas that urban policy was consolidating, thus, was closely linked to local conditions without yet 'objective' criteria to determine which neighbourhoods to include. In this sense, it was a 'local geography', based on a knowledge of local conditions, and sensitive to local specificities (Estèbe, 2001). In a sense, the relatively small number of intervention areas made this kind of approach possible. After all, the DSQ was to be experimental and not necessarily permanent. But the geography of urban policy kept expanding. A year later, with the contracts signed between the state and regions in the framework of the IXth Plan (1984–8),[23] the number of urban policy neighbourhoods increased from 23 to 148, including now 1.5 million inhabitants and some 350,000 housing units (Figure 3.6).

When Jacques Chirac became Prime Minister in 1986 in the 'cohabitation' government with the Socialists,[24] the future of DSQ, as a programme initiated by the Socialists, was not clear. However, Pierre Méhaignerie, his centrist minister, pushed the government to maintain the programme in order to address the issue of immigration and to counter the rise in the popularity of the extreme right in the peripheral areas of cities (Collovald, 2001; Damamme and Jobert, 1995), although Chirac's party, the RPR, had once again benefited largely from local alliances with the FN in the 1986 elections (following which the FN entered the parliament with 35 members).

Urban policy, therefore, continued during the two years of the Chirac government. A report commissioned by Chirac was published in 1988,

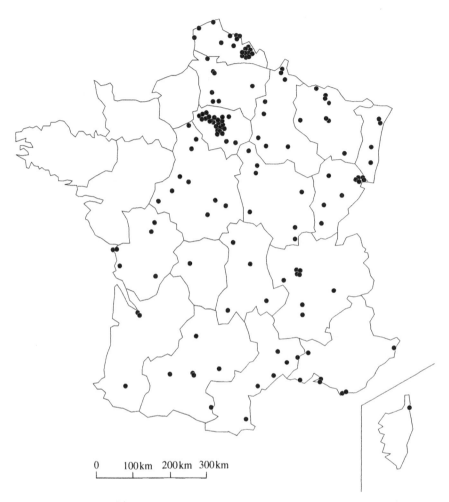

Figure 3.6   The 148 urban policy neighbourhoods (1984–8) (*Source: Délégation Interministérielle à la Ville* (DIV))

known as the Levy Report. The report's evaluation of the DSQ was largely positive, and the programme was seen as a 'national priority', the termination of which would have severe consequences (Levy Report, 1988: 63). The Levy Report, thus, opted for the continuation of urban policy, and added that the 'exceptional' quality of the DSQ had to be maintained. In this sense, it considered 150 to be a reasonable number for neighbourhoods to be included in the programme. However, the choice of these neighbourhoods had to be made more rigorously than it had been (the report called for 'a precise diagnostic'), and the programme had to be reserved for social housing neighbourhoods with *grands ensembles* in the peripheral areas of cities. These two propositions of the Levy Report were significant in that they diverged from earlier approaches in two ways. First, they proposed a shift from a more locally based selection process to a more 'objective' and centralized one. Second, they proposed a shift in the 'target' of DSQ interventions. Dubedout was in favour of not limiting the interventions to peripheral areas only, although most of the selected neighbourhoods were in the *banlieues*. The proposal of the Levy Report would make the DSQ a policy for the social housing neighbourhoods in *banlieues*.

As we will see in the next chapter, the Levy Report's proposal to devise criteria for the selection of urban policy neighbourhoods would be followed. However, as we will see shortly, its proposal to maintain the exceptional quality of urban policy by limiting the maximum number of included neighbourhoods to 150 would look almost bizarre compared to what urban policy became a year later. In fact, it would have been impossible not to have 'objective' and centralized criteria given the form urban policy took at the end of the 1980s – institutionalized, bureaucratized, consolidating and intervening upon an ever-expanding geography of neighbourhoods.

These transformations of urban policy took place during the second term of François Mitterrand as President, which started in 1988 with ambitious projects for the city. He declared that the city would be one of his priorities during his second term. In a message sent to a conference organized by *Banlieues 89*, Mitterrand qualified the work undertaken as 'inventing a real urban civilization', stated that he had always been supportive of the efforts that focused 'on the situation of certain *banlieues*' since 1981, and expressed the government's support in order 'to break the circle of degradation and despair, restore civility, and develop citizenship' (Banlieues 89, 1989: 3).

A few months following the start of Mitterrand's second term, which was no longer a cohabitation, Prime Minister Michel Rocard (PS) advanced reforms for the institutionalization of urban policy, which was part of the government's programme on the reform and renewal of public service (Le Galès, 2005). Indeed, the Rocard government's (1988–91) reform programme has been referred to as 'perhaps the most ambitious programme of public-sector reform attempted under the Fifth republic' (Knapp and Wright, 2001: 294). The decree of 28 October 1988 established the National Council of Cities (CNV, *Conseil National des Villes*), bringing together the CNDSQ and CNPD. It also created an inter-ministerial committee (CIV, *Comité Interministériel des Villes*), and the *Délégation Interministérielle à la Ville* (DIV). Thus, in 1988, the current institutional structure of urban policy was consolidated. The CIV, chaired by the Prime Minister, was charged with decision-making, the CNV with research and proposals, and the DIV with the coordination of actors and actions concerning urban policy. The creation of the DIV not only served to bring an administrative focus to urban policy programmes within other state institutions and policies, it also made clear that 'urban policy was given greater prominence, thereby symbolizing the commitment of the state to tackling the urban crisis and in turn encouraging a greater involvement of the various ministries' (Le Galès and Mawson, 1994: 27–8). However, this institutional structure also meant, as we will see in the following chapter, a departure from urban policy's experimental and militant approach towards a much more bureaucratic one.

A new programme started experimentally in 1989 with the preparation of 13 City Contracts (*Contrats de Ville*). These contracts defined a programme of action between the state and the localities for a period of five years, which was more comprehensive than the DSQ. This programme was conceived with three major objectives. The first was to change the scale of intervention from individual neighbourhoods to the entire city-region. Related to this, the second objective was to encourage mayors to take into consideration broader social and economic issues. And finally, it was hoped that this programme would foster inter-communal cooperation by bringing communes together in devising projects to address deprived areas. However, the City Contracts programme encountered serious problems in implementation. It was argued that the proposed projects failed to take into consideration deprived areas, and that some localities used the contracts as an opportunity to finance projects that had little or nothing to do with the so-called 'neighbourhoods in difficulty'. Inter-communal cooperation also was far behind expectations (Donzelot and Estèbe, 1999; Le Galès and Mawson, 1994; OECD, 1998).

In addition to institutionalization and the launch of the new City Contracts programme, there were two discursive changes in urban policy. The first concerned the title: Social Development of Neighbourhoods (DSQ) became Urban Social Development (DSU, *Développement Social Urbain*), implying a more comprehensive approach by changing the scale of intervention from the neighbourhood to the city. The second change concerned the name of the problem urban policy was supposed to address: (social) exclusion. Whether or not this change reflected European Union policies is open to debate, although the EU's emphasis on social exclusion in the 1990s was probably influential for the term's widespread use. As Percy-Smith, a policy analyst, wrote:

> The term 'social exclusion' originated in the social policy of French socialist governments of the 1980s and was used to refer to a disparate group of people living on the margins of society and, in particular, without access to the system of social insurance. [. . .] However, when the term began to be used

in the European context it referred more to the European Union objective of achieving social and economic cohesion. (2000: 1)

The turning point, according to Percy-Smith, was 1994, when the Council of Europe rejected a new poverty programme. Since then, she argues, 'social exclusion' rather than 'poverty' has become the major focus of EU social policy. In France, the term 'exclusion' first entered the agenda in 1974 with the publications of René Lenoir's *Les exclus*. Paugam (1996), in his account of the evolution of this term, argues that it was only in the late 1980s that the term re-appeared in political discourse in France. This had to do with the passing of a new law (with a unanimous vote in the National Assembly) in 1988, the *Revenu Minimum d'Insertion* (Minimum Insertion Income, RMI hereafter).[25] The RMI offered a minimum income to unemployed people over the age of 25 provided that they were not students, and were willing to be trained for or placed in work. This highly publicized and largely supported law was presented as a fight against (social) exclusion, which was an important part of the Socialist government's discourse at the time. As we will see in the next chapter, the term 'exclusion' became an integral part of urban policy in the 1990s.

A final change in urban policy in the late 1980s concerned the number of neighbourhoods included in the programme, which has continued to increase since the policy's inception. With the Xth Plan (1989–93), the number of neighbourhoods went up from 148 to 400 (Figure 3.7). Urban policy was no longer an experimental policy concerned with a few neighbourhoods, as Dubedout insisted, but rather an ambitious programme with 400 neighbourhoods to be tackled. The new list of urban policy neighbourhoods included the same neighbourhoods that were designated in 1982 (16 neighbourhoods), 1983 (23 neighbourhoods) and 1984 (148 neighbourhoods). In other words, all the neighbourhoods that were included in urban policy since its inception in 1982 were among the 400 neighbourhoods of 1989. Once designated, it seemed, there was no way out.

Conclusions: Consolidation of the Police

The early years of urban policy were marked by an aspiration to initiate a local political dynamic in its selected neighbourhoods by building upon their specificities – a reflection of the ideological affiliation of its initiators to *autogestion* and urban movements of the 1960s and 1970s. However, the initial political ideals of urban policy found little realization. Following the so-called 'anti-immigrant vote' of the 1983 municipal elections, the political context in which the social housing neighbourhoods of *banlieues* were addressed changed significantly. In a context of rising economic insecurity,

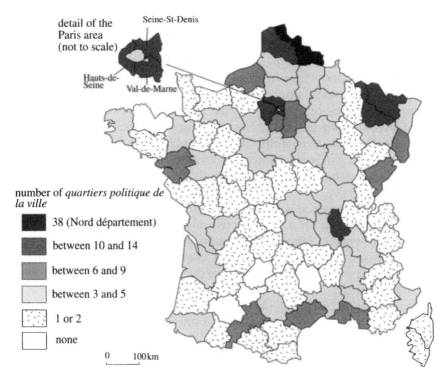

Figure 3.7 The 400 urban policy neighbourhoods per department (1989–93) (*Source: Délégation Interministérielle à la Ville* (DIV))

immigration and citizenship became salient issues in the 1980s. Social housing neighbourhoods in the *banlieues* spatially reified growing concerns about these issues, which were tightly linked to urban policy right from the start, already haunted as it was by the ghost of the so-called 'Anglo-Saxon world' unleashed by the Brixton revolts. There was a marked effort to avoid stigmatizing such areas with a lazy 'ghetto' label in the early years of urban policy. The presence of different cultures and ways of life were seen as a condition of urban life, which had to be recognized and respected. The problems of peripheral social housing neighbourhoods were related to the negative consequences of the economic crisis settling in (rather than, say, to 'cultural differences'), and such areas were referred to as working-class neighbourhoods – the ones most hard-hit by the loss of industrial and manufacturing jobs. In other words, there was an attempt to relate the *banlieues* to larger processes rather than articulating them as a 'threat' in and of themselves.

The initiators of urban policy saw a spatial approach – as opposed to sectoral policies – as an effective one in pursuing their political ideals, from *autogestion* to right to the city. The spatial designations were driven more by a desire to foster such political ideals than by a concern to institute a governmental order. There were concerns with stigmatization, which had already led some mayors to refuse inclusion in urban policy. But a choice was made in favour of a spatial approach, with the conviction that unless efforts were spatially focused on social housing neighbourhoods, it would always be the other parts of the city that would benefit from the measures (interviews with Sylvie Harburger and Pierre Saragoussi). 'Because it was always the [social housing] neighbourhoods that were forgotten' (interview, Sylvie Harburger).

The CNDSQ insisted, however, that urban policy be an experimental programme, focused on a few selected neighbourhoods with the aim of generating new ways of thinking politically about the city. It could generate ideas for future policies, but the initial programme was seen as far from permanent: 'We thought if it worked we ought to stop at that' (interview, Sylvie Harburger).

However, as urban policy institutionalized and the number of neighbourhoods increased (from an initial 16 to 400 in less than a decade), urban policy started to become a means for financing renovation projects and preventing revolts. In the meantime, the earlier emphasis on generating a local political dynamic started to disappear. Furthermore, once designated, it seemed unlikely that neighbourhoods could escape the geography of urban policy, which only continued to expand. The police order was starting to consolidate, making the social housing neighbourhoods of *banlieues* – some of which had been included for almost a decade now – the 'usual' objects of urban policy.

In terms of spatial conceptualizations, the geography of urban policy neighbourhoods in the 1980s was a 'local geography', defined through negotiations with local officials, paying particular attention to local dynamics. The selection of neighbourhoods was based on local knowledge and specificities rather than on centrally defined criteria. The spatially targeted areas were seen as part of their larger geographies, not as separate, self-contained areas.

This way of doing things was perhaps possible in an experimental policy dealing with a few selected areas, run mainly by a small number of commission members pursuing political ideals of a seemingly other epoch. But it just did not seem right for an ambitious urban policy preparing for the coming decade with new institutions, a new problem to tackle (i.e. exclusion) and a list of some 400 neighbourhoods. Urban policy in the new decade would have adequate institutions, a clearly defined problem to address and more precise and 'objective' knowledge of its ever more numerous intervention areas. But these reforms, as we will see in the next chapter, would also shift the orientation of urban policy from a political to a procedural one, proceeding with bureaucratic – and no longer militant – forms of intervention.

4

Justice, Police, Statistics: Surveillance of Spaces of Intervention

What would the young people of a 'sensitive neighbourhood' (one of the names given to the urban policy neighbourhoods), organized in the form of an association, have to do if they wanted to buy a ping-pong table?

1. Write out a consistent project report, underlining the educational value of the said table.
2. Get an estimate for the table.
3. Attach to their request the statutes of the association, the budget of the previous year, an updated list of the administrative council and the board.
4. Provide a deliberation of the council of administration confirming the firm intent on the part of the association to buy the table.
5. Place the application with the municipality for examination by the head of the DSQ project.
6. Wait until the application is transmitted to the local commission.
7. Then wait for the application to move on to the prefecture for submission to the technical committee for cities. Everybody then meets up and gives an opinion on the table: the sous-prefect, the regional council, the DRASS, the DRAC, the DRE, the DRFP, the DRPJJ, the DRJSP, the CAF, the National Education, the regional delegation of the FAS . . . Is the project valid? Is it really educational? Do all the administrations agree for this purchase? Is the city associated? Is the municipality's intent firm? After the table, what further educational content can be foreseen?[1]

A rather overlooked report was published in early 1990, the findings of which had inspired this ping-pong table joke at the time. The report was quite critical of the workings of urban policy, which, as we have seen in the previous chapter, was institutionalized along a more bureaucratic line at the end of the 1980s (hence the joke). As we will see in this chapter, this trend would continue in the early years of the 1990s, transforming urban policy from a militant to a bureaucratic form of intervention. This transformation would also affect the ways in which the spaces of urban policy were constituted; statistics would replace local knowledge and specificities, constituting the spaces of urban policy as clearly defined objects of intervention. Thus consolidated, this police order would also provide the basis for the engagement of the Ministry of Justice and the French Intelligence Service with the question of *banlieues*.

But let us go back to the report for the moment, which was prepared upon the request of Prime Minister Michel Rocard following the incidents of unrest in Chanteloup-les-Vignes, a *banlieue* of Paris.[2] The author of the report, Claude Sardais, was quite critical of the implementation of urban policy, which he was asked to evaluate. The concern of the Prime Minister was understandable: despite the state's engagement with the question of *banlieues*, incidents of unrest recurred. Chanteloup-les-Vignes had been included in urban policy since 1983; it was, in other words, one of the 'priority neighbourhoods' of urban policy for seven years at the time of the incidents.

The Sardais Report (1990), as I mentioned, received relatively little attention (though it did inspire the ping-pong table joke). There seem to be two possible explanations for this. The first is that it was not really overlooked, but was simply overshadowed by the incidents of Vaulx-en-Velin, which occurred a few months after Chanteloup-les-Vignes, and received unprecedented media and political attention (as we will see shortly). The second has more to do with the content of the report than with the context in which it was prepared; it was, perhaps, Sardais's rather severe critique that contributed to the relative neglect of the report. Sardais was a member of the Financial Inspectorate (*Inspection des Finances*), one of the most prestigious institutions of French higher administration. Being an *Inspector des Finances*, he was mainly focused on the workings of urban policy in terms of credits, although, interestingly enough, his report was structured around the themes of citizenship and recognition. We will also see that his critique of urban policy resonates well with the critique made by the local activists in Vaulx-en-Velin, which will be presented in Chapter 6. Indeed, in an annex to the report written shortly after the incidents of Vaulx-en-Velin, Sardais wrote the following:

> In Vaulx-en-Velin, it is the policy implemented that is to be blamed: the major part of the means is spent on rehabilitating and improving the built environment while the social and educational approach is merely palliative. This conception is insufficient to deal with the anxiety of adolescents and the

young adults faced with failure at school and lack of stable professional insertion. (Annex to Sardais Report, 1990: 1)

One of the Sardais Report's findings was that the state delivered the
DSQ money with long delays, which undermined both the efficiency and
the credibility of the programme. Waiting periods for the allocation of state
money were too long, procedures too complicated, and structures too
inconsistent. Furthermore, Sardais argued, the realizations of projects were
badly coordinated, they did not correspond to the priorities of the programme, and the opinion of inhabitants was hardly taken into account. In
Chanteloup-les-Vignes, for example, the local association was attributed a
subvention for the year 1989, which was still not delivered in 1990. Indeed,
the issue was not yet even on the agenda of the Municipal Council. Only
after the incidents was the money delivered, following the intervention of
higher authorities. Sardais (1990: 23) wrote: 'How not to conclude that
"the administration only understands the language of violence", as the
author of this report heard several times from disillusioned local actors?'
But the problem, for Sardais, was not simply a matter of a system that
worked too slowly. Certain groups of the population seemed to be systematically left out of political decisions, and even 'denied existence':

These [social] categories bordering on exclusion (on either side of the border)
weigh little, or even not at all in the case of immigrants, from an electoral
point of view; they have no weight from the economic point of view (they do
not possess capital and are not organized in unions), they are, most of them,
denied existence from the cultural point of view, they are not organized as a
lobby group: they do not exist in media reports, except for events of the type
of what happened in Les Minguettes or Chanteloup-les-Vignes. But do they
exist? That is the worrying question the author of this report felt amongst
many of the young people he met in the cités and who have a strong desire
for recognition. (1990: 68)

Sardais, therefore, argued that the issue was beyond simply 'fine-tuning' a
system that worked in an unsatisfactory way. There was a strong desire for
recognition in the cités, the denial of which seemed to contribute to the
recurrence of such incidents. The system, yes, was not working efficiently,
but the problem was more profound than that. As long as the young people
in these cités were systematically denied recognition and rejected from
the school system and the job market, Sardais argued, 'claiming their mar-
ginality' would remain the only way for them to construct their identity
(1990: 126).

For Sardais, the goals of urban policy were quite ambiguous. Furthermore, the extraordinary mediatization of urban policy posed problems.

Some mayors, Sardais argued, might be too busy trying to get subventions rather than really engaging with the economic *and* social development of neighbourhoods. The whole system, he argued, was aimed more at 'financial visibility' than at 'action and implementation' (1990: 44). He proposed, therefore, that the main line of urban policy be clearly defined, taking into consideration the constant exclusion of certain groups from the policy process. His proposal was to focus urban policy actions on 'the insertion of inhabitants through economic and social development of disadvantaged neighbourhoods and *banlieues*, [and] the construction of their citizenship with an emphasis on the young population' (1990: 90). He held that providing the necessary conditions for the inhabitants to fully enjoy their citizenship was as important as improving material conditions, and insisted that insertion had to been seen as at once economic, social and civic.

The conceptualization of urban policy was no less problematic for Sardais. When urban policy was first conceived, Dubedout had insisted on the 'exceptional' quality of the programme with a small number of sites to be included. However, the number of sites (i.e. 'priority neighbourhoods' of urban policy) has constantly increased since then, rising, as we saw in the previous chapter, from an initial 16 to 400 in less than a decade, which, according to Sardais, resulted in the loss of originality of the state's intervention. What started as an innovative and political spatial approach in the early 1980s became a relatively stable spatial order for intervention – more procedural and bureaucratic – in the early 1990s.

The incidents of Chanteloup-les-Vignes and the Sardais Report, however, did not receive much attention. The decisive moment was still yet to come, a few months later, in Vaulx-en-Velin.

When the Margin is at the Centre

Was it a marginal problem, that is, on the margins of the city, on the margins of institutions? Or was it a central problem that challenged the very evolution of French society, through the issue of its cities and its cohesion? Public opinion, as well as most elected officials and professionals in the early 1980s, were convinced that it was a marginal problem. In 1989 [*sic*],[3] after the incidents of Vaulx-en-Velin, which demonstrated the frailty of the urban balance, the message was different, and the discourse was reversed. The depth of the crisis had transformed what was only an exception into a central problem.

Harburger, 1994: 388

The weekend of 6–7 October 1990 came as a big surprise to the government and local officials: Vaulx-en-Velin, which was until then seen as one of the 'exemplary' sites of urban policy, included in the programme since 1984, was in flames. Incidents started on the 6th, following the death (killing, according to the young people of the neighbourhood; accident, according to the police) of a young person of immigrant origin by a police car, and continued most intensely for two days. The reverberation of incidents went well beyond urban policy, and Vaulx-en-Velin immediately entered the national political agenda. Remarks from President Mitterrand and Prime Minister Rocard, visits to the site by urban policy authorities and other officials, and a discussion at the National Assembly immediately followed the incidents. Politicians of all colours, 'experts', academics, journalists – everyone had something to say, a remark to make, a message to pass on, as if everyone was waiting for Vaulx-en-Velin to happen. For some reason, Vaulx-en-Velin was not 'just another site' of urban policy; it stood for something more.[4] As Boubeker (1997: 87–8) wrote, with the incidents of Vaulx-en-Velin,

> the public images of the *banlieue* and of immigration allow French society to discover its internal frontiers. The malaise of *banlieues* puts into perspective the failures of the republican model of integration and the decay of the abstract universality of a community of citizens: the unity of the social body seems to be challenged by the emptiness of political response to urgent cultural and social claims which remain unsatisfied.

But why did Vaulx-en-Velin stand for something more? One of the reasons, apparently, was that it had been seen as an exemplary site of urban policy since 1984. And in this sense, it was the first time that urban policy was being contested (interview, Sylvie Harburger). With the incidents, Vaulx-en-Velin's reputation changed from being an exemplary site of urban policy to one of its 'two founding events' (CNV, 1991b: 11), next to Les Minguettes, remembered for the incidents of the 'hot summer' of 1981. But compared to Vaulx-en-Velin, Les Minguettes seemed rather trivial in terms of the mediatization of incidents. The change brought about by the incidents of Vaulx-en-Velin was both quantitative and qualitative, as a press review made clear (CNV, 1991b). Quantitatively, there was a genuine explosion in the number of experts and specialists (mostly sociologists) appearing in the newspapers presenting their views on what had happened in Vaulx-en-Velin. Moreover, some journalists were now specialized in the issue of *banlieues*. By contrast, nine years earlier, following the incidents of Les Minguettes, there was merely one person interviewed by a newspaper, Bernard Grasset, the then prefect of the department of Rhône, and the attention of the media was remarkably low. In the major national newspapers, for example, only a little paragraph had appeared in *Le Monde*, *Le Figaro* and *L'Humanité*, and nothing in *Libération*. In the week following the incidents of Vaulx-en-Velin, however, there were some 60 articles, nine editorials and 34 reports diffused on the radio or television (Bonelli, 2001). *Libération*, which had passed Les Minguettes in silence, this time published a special section on Vaulx-en-Velin, and devoted its cover page to the incidents, with the title 'Why Vaulx-en-Velin?' (13–14 October 1990). Qualitatively, interviews, information boxes and maps were presented when reporting the incidents, presenting a mix of journalistic accounts, commentary, the expert's view and the specialists' comments. All of these contributed to the emergence of new actors and new discourses concerning the *banlieues*. Indeed, as Hargreaves (1996) shows, the construction of the *banlieues* as a journalistic category followed the incidents of Vaulx-en-Velin. As we will see in this chapter, the incidents also led to the engagement of other state institutions – namely, the Ministry of Justice and the French Intelligence Service – with the question of *banlieues*, contributing to the articulation of *banlieues* more in terms of 'threat'.

The exemplarity of Vaulx-en-Velin prior to the revolts and the magnitude of incidents, however, only partly accounts for the unprecedented media and political attention. A crucial factor that made Vaulx-en-Velin a turning point was the context in which the incidents occurred. Like Les Minguettes in 1981, the incidents of 1990 in Vaulx-en-Velin occurred in a context of hotly debated issues, national as well as international, which resonated with the anxieties at home about immigration and Islam. The Intifada had already been in progress for three years (which led to comments about the 'Intifada of the *banlieues*'), the Salman Rushdie affair had happened only a couple of years previously, and the Gulf War was about to start. The early 1990s was also particular in that the continent was no longer divided neatly to distinguish friend from enemy following the demise of the Berlin Wall in 1989, urging the French state, already preoccupied with 'menaces to French identity', to affirm its authority and sovereignty in the midst of political restructuring that modified seemingly immutable political spaces. In addition to these developments, the Los Angeles riots occurred two years after Vaulx-en-Velin, summoning once again the ghost of the 'ethnic' nightmare haunting the French republic. In France, there was a hot debate following the Islamic headscarf affair of 1989, to which the 'public response was almost unanimously hostile, not to say at times hysterical' (Jennings, 2000: 584).[5]

The reflection of these events in France was a resurgence of debates around, and arguments against, immigrants and 'communities', focusing particularly on North African immigrants and Islam.[6] Just as the context in which the incidents of Les Minguettes occurred led to the articulation of *banlieues* with immigration, the context in which the incidents of Vaulx-en-Velin occurred led to their articulation with Islam – or, better yet, with a particular interpretation of Islam seen to be incompatible with the values of the republic, and prone to taking fundamentalist and terrorist turns. Islam in *banlieues* was not a novelty. Indeed, as Battegay and Boubeker (1991–2: 58) argue, it was until then seen as a 'good thing' by the administrators, who thought it would calm down the *banlieue* youth a bit. In the context of the late 1980s and early 1990s, however, the issue of Islam entered the debate around *banlieues*, following the incidents in Vaulx-en-Velin and other *banlieues* in the following years.[7] To this was added Islamic terrorism following the bomb attacks in Paris in 1995, perpetrated by a group of second-generation Algerians, one of whom – Khaled Kelkal from Vaulx-en-Velin – was shot dead by the police. Khaled became an emblematic figure of the articulation *banlieue*–Islamic terrorism, not least because his killing was screened on prime-time television only minutes after the event – where, as it turned out, an off-camera police voice calling 'Finish him!' while he lay wounded on the ground was censored.[8]

But incidents were not confined merely to peripheral areas. In Paris, high-school students were on the streets a month after Vaulx-en-Velin, protesting against the projects of then Minister of Education, Lionel Jospin. The manifestations, however, produced unintended consequences: shop windows were smashed in Montparnasse, a fairly well-off neighbourhood of Paris, followed by looting of clothes from stores and confrontations with the police. The dominant discourse in the media was that the young people doing the damage were coming from the *banlieues*, and were mainly of North African and African origin. This conviction was not limited to the media only. A telling example is a report presented two years later at the National Assembly by Julien Dray, one of the co-founders of the anti-racist organization *SOS-Racisme* and now a member of the Socialist Party. His report was entitled *Violence of Young People in the Banlieues*, the idea for which was 'born out of a shock during the demonstrations by high-school students in the Fall of 1989 [*sic*]' (Dray, 1992: 9). In other words, what prompted Dray to present a report on the 'violence of young people in the *banlieues*' were incidents that took place not in the *banlieues* but in the middle of Paris (which, in fact, took place not in 1989 but in 1990). *Banlieues* were now coming into the city.

The National Council of Cities (CNV) attempted to counter this image with a study published shortly after the incidents in order to find out who these young people really were. The report looked at the nationalities of 46 young people questioned by the police, stating that most of them (31 out of 46) were in fact French citizens.[9]Although a high proportion of *banlieue* youth were implicated in the incidents, the report stated, there was no sign of a premeditated mobilization in the *banlieues* to come and provoke incidents in Paris, except, perhaps, in a few cases (CNV, 1990a). *Banlieues*, once again, were at the centre of the political agenda. The French state, as it did in 1981, took the issue seriously.

The 'Return of the State'

The early 1990s, as far as urban policy goes, has been referred to as 'the return of the state' (Barthélémy, 1995; CNV, 1992a; Merlin, 1998), with the passing of new laws, the creation of new institutional structures and a proliferation of reports in order to 'fight against exclusion' and to 'prevent ghettoization' of certain neighbourhoods. The so-called 'return of the state' was first implied by President Mitterrand in his frequently cited speech on 4 December 1990 at Bron, a *banlieue* of Lyon just south of Vaulx-en-Velin (see Figure 3.4). Mitterrand stated that it was 'possible to design, in this hideous jumble [*magma*] of *banlieues* of big cities, an order, an urbanism, an aesthetics, a way of life, and maybe a hope' (Mitterrand, 1990: 2).

Mitterrand proposed four main lines of action. The first was to concentrate the efforts of the state on the 400 urban policy neighbourhoods, now referred to as 'neighbourhoods in difficulty'. The second had to do with the diversification of activities in these neighbourhoods, which were seen to be suffering from, among other things, architectural uniformity (*grands ensembles*) and separation of functions. The third was the participation of inhabitants – a notion that Cubero (2002–3) referred to as the 'eternal *leitmotiv*' of urban policy. The last one was the creation of jobs for the inhabitants of these neighbourhoods, which had remained a major problem since the early 1980s.

Mitterrand also proposed the assignment of sous-prefects to certain departments, who would be in charge of the neighbourhoods in question as local arms of the state, and evoked the possibility of creating a Ministry for the City. The assignment of sous-prefects as local arms of the state seemed contradictory compared to the reforms realized during Mitterrand's first term. With the decentralization reforms of the 1980s, the centrally appointed prefect had lost *a priori* supervisory control over the communes, and his executive powers over departments and regions were transferred to the elected presidents of departmental and regional councils. These changes were not merely administrative; they also carried symbolic power, even though the prefect still remained the representative of the central state. While Mitterrand's proposal did not imply a transfer of all the local power to the sous-prefect, the assignment was still made by the central state. As Nakano (2000: 98) argues, '*Monsieur le préfet* was once the very embodiment of the centralising, overbearing state in France.' The decentralization reforms of the early 1980s, in this sense, reflected the commitment of the Socialist government to locally elected officials, rather than to centrally appointed representatives of the state, which was also evident in urban policy by the appointment of Dubedout. With the reforms, even the title was changed from 'prefect' to '*Commissaire de la République*'.[10] As we will see in the following chapter, the prefects' role in urban policy would be reinforced in the 1990s with the nomination of a prefect to the head of the DIV and the initiation of a major urban policy programme conceived by two prefects.

Two weeks after Mitterrand's Bron speech, on 21 December 1990, Michel Delebarre (mayor of Dunkerque) was appointed as the Minister of State for the City, and the following month, 13 sous-prefects were assigned to the 'most sensitive' departments, charged with urban policy as their mission.[11] One of the missions of the new Minister for the City was 'eliminating exclusion' (DIV, 1999: 8).

In addition to new institutionalizations, the 'return of the state' of the early 1990s in the domain of urban policy was marked by the passing of new laws. The 'Besson Law' of 31 May 1990 defined the right to housing (*droit au logement*) as a 'duty of solidarity for the entire nation', and required the departments to financially contribute to 'solidarity funds' at least as much as did the state (Articles 6 and 7). More importantly perhaps, it prohibited the use of the pre-emptive right against social housing projects in communes where social housing accounted for less than 20% of housing (Article 14). This was an important measure since some mayors used their pre-emptive right whenever a social housing organization was looking for land in their communes (see, for example, *Libération*, 13 July 2006).

Another law on 'financial solidarity' – the *Loi de dotation de solidarité urbaine* of 13 May 1991 – was aimed at establishing inter-communal solidarity (as was suggested by the Dubedout Report of 1983 and Pesce report of 1984) through a transfer of funds from richer communes with fewer social problems to poorer communes with more social problems. This was a significant redistributive measure that was aimed at fostering solidarity between communes.[12]

The most mediatized of the new laws, however, was the *Loi d'Orientation pour la Ville* (LOV) of 13 July 1991, known also as the 'anti-ghetto law'. The LOV was presented at the National Assembly in April 1991 by the new Minister for the City. In his presentation, Delebarre identified segregation and social exclusion as major problems, induced partly by the concentration of populations with difficulties in certain areas. The new law was conceived to address this problem by encouraging the dispersal of future social housing construction. What was at stake, Delebarre stated, was 'the cohesion of French society', and the aim was 'restoring cohesion' (Projet de loi, 24 avril 1991: 4). The LOV was published in the official journal on 19 July 1991, and its first article read:

> In order to realize *the right to the city*, urban districts, other territorial collectivities and their groupings, the state and its public institutions assure to all the inhabitants of cities conditions of living and dwelling promoting social

cohesion as to avoid or abate the phenomena of segregation. This policy must provide for the insertion of each neighbourhood in the city and assure the coexistence of diverse social categories in each agglomeration.

The reverberation of Lefebvre's notion of the right to the city was clear:

> To exclude the *urban* from groups, classes, individuals, is also to exclude them from civilization, if from not society itself. The *right to the city* legitimates the refusal to allow oneself to be removed from urban reality by a discriminatory and segregative organization. (Lefebvre, 1996: 195)

The LOV was basically concerned with social housing. It was designed to oblige communes located in agglomerations with more than 200,000 inhabitants to provide 20% social housing. The aim was to avoid further concentration of social housing in agglomerations that already had a high proportion of it, and to achieve 'social mixity' in agglomerations that did not have much social housing. Although the idea of a fairer distribution of social housing among agglomerations was a significant step, the LOV nevertheless seemed far from achieving its stated objectives – 'social cohesion' and 'the right to the city'. Concerning the former, the law was premised upon a supposed correlation between residential mixity and social cohesion (Simon, 1995). Concerning the latter, it transformed Lefebvre's political notion of the right to the city into a procedural one – which conformed to the bureaucratic shift in urban policy in the early 1990s. But even procedurally it was not clear what the notion really implied. A letter, then, was circulated on 31 July 1991 to clarify the articles of the law, which stated that the opening article on the right to the city had 'no normative nature'. Why was it there then? As 'a homage to the work of Lefebvre', perhaps, as Véronique de Rudder put it.[13]

Both the LOV and the 'Besson Law', however, remained largely ineffective due to 'bureaucratic inertia' (Merlin, 1998: 148). A year after the passing of LOV, for example, only two decrees – out of more than a dozen necessary for its implementation – had been published. Furthermore, the obligation for the mayors to organize meetings for each urban project that influenced the inhabitants was relinquished under the pretext of giving 'free rein to local initiatives' (*Le Monde*, 21 July 1992: 9).[14] Finally, there was resistance on the side of agglomerations to provide social housing. In this respect, the LOV was not assertive enough, although this issue was brought to attention by the *Conseil économique et social* (1991) during the preparation of the law. The *Conseil* stated that the law did not push hard enough in terms of solidarity between communes, but 'simply expect[ed] that each commune [would] accept and organise the mixity of populations' (1991: 6). The LOV was, thus, never fully implemented. The necessary decrees for its implementation were still not completely prepared when the Balladur government took office in 1993 and modified it with the passing of the so-called 'Carrez Law' in 1995, which largely eliminated the measures that sought to achieve a more balanced distribution of social housing.[15]

The institutional reforms and the new laws of the early 1990s were nevertheless representative of the state's commitment to the urban question. It was also necessary to define the orientations of urban policy for the new decade. In his mission letter dated 19 February 1991, Delebarre defined the priority of urban policy as 'the insertion of disadvantaged neighbourhoods and populations in the city', and requested a report from Jean-Marie Delarue, who became the head of DIV the same year.

'I Like the State'

> *Interviewer:* Bourdieu talks about technocratic arrogance combined with political marketing to characterize state policy.
> *Jean-Marie Delarue:* I like the state.
>
> *Panoramiques* 1993: 157

Jean-Marie Delarue's report, *Banlieues en difficultés: La relégation* (*Banlieues* in difficulty: the relegation), was published in 1991. Previous reports defining the orientations of urban policy were prepared by mayors with a knowledge and experience of local conditions (with the exception of Schwartz, who was an academic, and Levy and Sardais, whose reports were aimed more at evaluation than at defining the course of urban policy). Delarue's profile differed remarkably: a graduate of the *Ecole Nationale d'Administration*, the elite school for the civil servants of French administration, he was counsel of the *Conseil d'Etat*, the highest administrative court in France. His proposals for urban policy did not break new ground compared to those in the earlier reports. He was, nevertheless, more focused on the workings of the system than on promoting a local political dynamic, as the initiators of urban policy were. In this sense, the Delarue Report marks a turning point in urban policy – which is perhaps best captured by the epigraph that opens this section – towards a better institutionalized, and more bureaucratic and technocratic approach.

The Delarue Report is also emblematic of a change in urban policy discourse in the 1990s, not so much for its literary allusions as for its reference to 'the republican tradition' for the first time in an urban policy document. The report was structured around the notion of citizenship, which had to be reaffirmed in the 'relegated' neighbourhoods. Otherwise, Delarue maintained, there would be 'alternative solutions', some of which could 'already be discerned'. These included '"communitarian" groupings, alien to the republican tradition, as emphasized in its first report by the *Haut Conseil à l'Intégration* [HCI]'; 'a Sicilian-style degradation in the *cités*'; and 'an extension of phenomena that are appearing marginally, these lawless [*sans foi ni loi*] places in which nothing and no one can take hold' (1991: 90).[16]

The Delarue Report's reference to the HCI, the republican tradition and communitarian groupings incompatible with it is significant. The HCI was created immediately after the Islamic headscarf affair of 1989 as council for advising the government on the issue of integration. The political agenda in the early 1990s was largely dominated by this issue – not 'integration' in general, but the 'integration' of non-European immigrants and their descendants into French society, which were the main 'targets' of the so-called 'republican model of integration', in practice if not in theory.[17] The HCI's aim was to conceive a 'republican model of integration', which eventually led, as de Rudder (2001: 30) argues, to 'an exclusive republican nationalism' with connotations of French identity and the acceptance of French cultural values. Although many scholars have noted the effects of republican nationalism on immigration and citizenship policies (see, for example, Balibar, 2001; Blatt, 1997; Feldblum, 1999), how it affected urban policy did not receive attention. As we will see, urban policy in the 1990s increasingly framed issues with references to 'the republic', and the Delarue Report, in this sense, is a turning point in urban policy discourse.

The Delarue Report shared the same conviction in terms of the role of urban policy programmes with the Levy Report. In 1988, the Levy Report had maintained that the situation would have been worse in the absence of such programmes. It was followed, two years later, by the incidents of Chanteloup-les-Vignes and Vaulx-en-Velin, the former in the seventh year of its inclusion in urban policy, and the latter in its sixth. Likewise, the Delarue Report held that urban policy programmes had played an important 'preventive' role in the selected neighbourhoods. It was followed by seven revolts in the same year, and four others in the following year. The first seven in 1991 took place in the social housing neighbourhoods of Sartrouville (included since 1989), Mantes-la-Jolie (included since 1982), Meaux (included since 1984), Garges-les-Gonesse (included since 1984), Aulnay-sous-Bois (included since 1984), Amiens (included since 1983) and Vénissieux (included since 1982). The four others in 1992 took place in Vaulx-en-Velin (included since 1984), Reims (included since 1984), Tourcoing (included since 1989) and Brunoy (later included in 1996).

This is a curious situation that suggests two non-exclusive possibilities (other than the possibility that this is pure coincidence). The first one is that despite being included in urban policy programmes for several years, these neighbourhoods suffered from embedded problems to which urban policy was unable to respond adequately. The other is that the actions carried out in the framework of urban policy programmes were conceived poorly, increasing the already latent tensions between young people and figures of authority, such as the municipality and the police. Indeed, the Vaulx-en-Velin story presented in Chapter 6 suggests such a possibility. One thing, however, was sure: revolts in the social housing neighbourhoods of *banlieues* recurred with increasing magnitude, and they expanded geographically. A working group of the National Council of Cities published a report in order to examine the 'movements of violence' that had occurred in October 1990 (CNV, 1991a, known also as the Cardo Report since the group was chaired by Pierre Cardo, mayor of Chanteloup-les-Vignes). This fairly short report stated that the majority of the young people implicated in these incidents were from immigrant backgrounds and came from socioeconomically disadvantaged families. Despite an element of violence, such incidents, it was argued, were nevertheless responses to a 'feeling of exclusion' generated by school failure, unemployment and lack of facilities in the *banlieues*. These young people, the report maintained, 'want a fairer society', and even if there had been recourse to violence during the incidents, the aim was to 'assert their claims' (CNV, 1991a: 4).

These remarks and the institutional reforms of the early 1990s show that revolts in the *banlieues* were still taken seriously by the government. However, the early 1990s, as noted above, was also a period in which other state institutions would engage with the question of *banlieues*, including the Ministry of Justice, the French Intelligence Service and the State Statistical Institute. These engagements, to put it in Rancière's terms, would introduce new sensible evidences (new categories, new statistics, new official terms and discourses – in short, new statements) and contribute to the further consolidation of the police order with a more precise delineation of the spaces of intervention.

Justice, Police, Statistics

In 1991, the Ministry of Justice published a document entitled 'The Law Acts in the City' (Ministère de la Justice, 1991).[18] After a reminder as to the two objectives of urban policy as social development of neighbourhoods and prevention of delinquency, it was stated that the 'law [was] mobilized to better respond to urban problems'. There was a concern with the weakening authority of the law in the city, and the aim was to have a 'more direct presence of the Law [*la Justice*] in *sensitive neighbourhoods*' – a new term entering the agenda in the early 1990s. The *Maisons de Justice et du Droit* (MJD) were conceived for this purpose. They were to bring the juridical system closer to inhabitants in order to 'reinforce the presence of the

Law' through a rapid, on the spot, treatment of delinquency in the so-called 'sensitive neighbourhoods' (Ministère de la Justice, 1991: 4 and 5). The following year, the Minister of Justice issued a circular, and stated that it was 'necessary to ensure that no criminal act, even minor, even committed by a young person, remain without judicial response. Greater swiftness in the punishment is also particularly important' (Circulaire no. 92–13 du 2 octobre 1992: 2).

The circular was not about penal policy in general, but about 'responses to *urban delinquency*' – another new term entering the agenda in the early 1990s. Thus, 'a new stage of prevention of delinquency' started in this period with the collaboration of the Ministry of Justice, the Ministry for the City and the DIV (Ministère de la Justice, 1991: 13). The spatial order consolidated by urban policy – 'the police' – was the starting point of this collaboration with the repressive machinery of the state. It also provided the spatial basis for the French Intelligence Service.

Another state institution that turned to urban issues in the early 1990s was the *Renseignements Généraux* (French Intelligence Service, RG hereafter). Following the incidents of Vaulx-en-Velin and the demonstrations by high-school students in Paris, a special section was created at the RG in December 1990. It was initially a sub-section of a larger unit at the RG, and was called 'Urban Violence' – another new term entering the agenda in the early 1990s. In April 1991, it became an autonomous section within the RG, and changed its name to 'Cities and *Banlieues*' (*Villes et banlieues*). Directly attached to the Ministry of the Interior, the section was, and still is, 'responsible for studying problems of violence in sensitive neighbourhoods' in collaboration with other actors, including those involved in urban policy. Lucienne Bui-Trong, a former high-school philosophy teacher, was the creator and head (until her retirement in 2002) of this section.[19]

> [O]ur mission was to have a general view on the issues of *banlieues*, on violence in the *banlieues*, we wanted to be able to give the Minister of the Interior an assessment of the situation, we wanted to be able to anticipate riots, to be prepared in case serious incidents occurred . . . and we also wanted to help all the police forces to deal with this problem, while of course working with the other actors, since there was also an urban policy operating, we tried to see how the police could work in harmony with the other actors. (Interview, Lucienne Bui-Trong)

With a focus mainly on the *banlieues*, the section started work analysing and anticipating 'urban violence' and terrorist activities:

> [W]e are not visible for the population. [. . .] We try to keep up with the evolution of opinions, we are a bit like meteorologists of opinion. Therefore

we work in two directions, in fact: first, information on the problems, facts
of society, of opinion, and, on the other hand, we also work against terrorism,
against terrorist violence, where we work in a more focused way. [. . .] In our
services, there were two ways of working, so, on the one hand, there was the
work in which you introduce yourself, state you're working for the prefect,
for the Ministry of Interior, and you gather information, and, on the other
hand, a more focused infiltration work, to anticipate on . . . on terrorism
itself. (Interview, Lucienne Bui-Trong)

With the creation of this section concerned with 'urban violence', the
RG moved in new directions. As Bonelli (2001) argues, the RG has traditionally been concerned with the defence of institutions and social order.
It was specialized in the surveillance of 'subversive' political and social
movements, which involved the surveillance of the activities of government
officials, prefects and politicians in general. Especially in the 1970s and
until the mid-1980s, it was largely preoccupied with the 'menace' of leftist
'political subversion'. But since then, such a 'menace' has largely disappeared. Furthermore, the surveillance of the activities of politicians created
a lot of tension. Therefore in the early 1990s, the RG lacked 'legitimate
targets' of its own. Indeed, even the organization's dissolution was envisaged. The notion of 'urban violence' came as a 'lifesaver' to the RG; it had
a collective dimension, and thus required the organization's involvement.
The urban policy neighbourhoods were taken as a starting point. In 1991,
800 'sensitive neighbourhoods' were under surveillance by the RG, which
went up to 1,200 by the end of the decade (Bui-Trong, 2000).

Since then, the activities of the 'Cities and *Banlieues*' section have become
highly mediatized, especially following the publication of the so-called
'Bui-Trong scale' of 'indicators of violence in sensitive neighbourhoods'
(Bui-Trong, 1993). The scale consisted of eight degrees, ranging from acts
of vandalism and delinquency with no institutional connotations (degree
1) to confrontations with the police, 'guerrilla' and 'riot' (degree 8). Using
this scale, 'a cartography of sensitive neighbourhoods' was prepared. As
Bui-Trong explains, 'the first practical interest of the cartography of sensitive neighbourhoods is that it allows to better target (qualitatively) the
efforts of urban policy' (1993: 245). Thus, the section started to collaborate
with the newly created Ministry for the City.

[M]y own criteria have nothing to do with those of urban policy. But what
happens is, and that's the issue, of course, *most riots we had happened, took
place in neighbourhoods which were included in urban policy*, so of course our
work interested, from that very moment, it interested the Ministry for the
City, so we sent our work to them too. But, you see, the fact that the neighbourhood was or was not included in urban policy was never a criterion for
us to get interested in it. (Interview, Lucienne Bui-Trong)

The data collected by the RG, including the list of neighbourhoods under surveillance, are confidential and not released. There is, however, evidence that suggests that the neighbourhoods under surveillance largely overlap with the urban policy neighbourhoods. All but two of the areas where large-scale (degree 8) revolts occurred in the 1990s are urban policy neighbourhoods.[20] Furthermore, a map prepared by Bui-Trong (without giving the names of neighbourhoods) suggests such an overlap. This map shows the number of neighbourhoods under surveillance by department (Bui-Trong, 2000: 132). A similar pattern occurs when a comparable map is prepared for the urban policy neighbourhoods (Figures 4.1 and 4.2)

This overlap is not very surprising because when it was decided, in 1990, that the RG would keep certain neighbourhoods under surveillance, it was the list of urban policy neighbourhoods that was directly taken as a starting point (Bui-Trong, 2000). The list included 546 neighbourhoods at the time. Since then, however, the RG has developed its own criteria, and the list expanded gradually, from 800 in 1991 to some 1,200 by the end of the decade.

While the data collected by the RG remain confidential, it is possible to obtain some information about these neighbourhoods by using another set of data provided by the State Statistical Institute (*Institut National de la Statistique et des Etudes Economiques*; INSEE, hereafter), the second body with which the Ministry for the City would collaborate in the early 1990s. The problem for the new ministry, whose mission was defined as 'fighting exclusion', was to spatially constitute its object of intervention: that is, spatially defining and delimiting 'exclusion'. Fighting exclusion meant to fight it spatially, given the way urban policy had operated since its inception. However, this had to be done with an *already* existing geography of priority neighbourhoods – the 'local geography' of the 1980s – which initially was to have 400 neighbourhoods, but consisted of 546 by the time INSEE got engaged. Thus the Ministry for the City started to collaborate with INSEE for the statistical profiling and mapping of the urban policy neighbourhoods. A section called 'City' (*Ville*) was created at INSEE to this purpose. This collaboration was aimed at re-constituting the spaces of urban policy as the objects of a national policy in a more rational and precise way. As Estèbe (2001: 31) put it: 'What is at stake is the transformation of the essentially local geography of the 1980s – based on reputations, empirical and grounded knowledge [*des connaissances sensibles et empiriques*], parallel histories of cities and neighbourhoods – into a set of territories likely to become an acceptable object for a national policy for fighting exclusion.' The neighbourhoods were thus re-constituted as objects of a national policy for spatially fighting exclusion, using long-term unemployment, young people under 25 and foreigners as criteria. This new geography of the priority neighbourhoods of urban policy was a 'relative

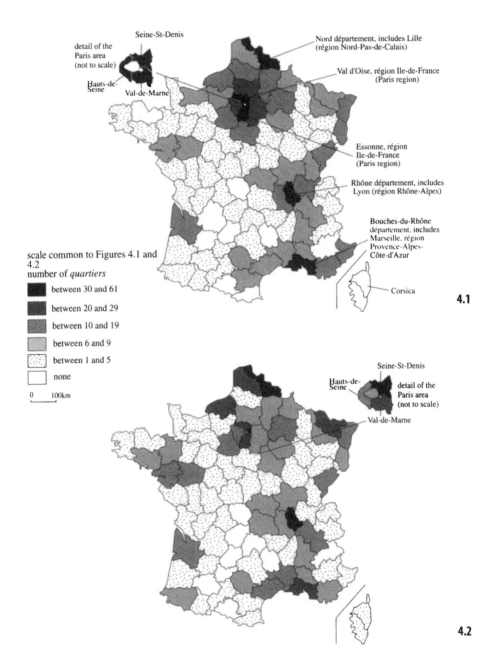

Figure 4.1   The 'difficult neighbourhoods' of the French Intelligence Service per department (1999) (*Source*: Adapted from Bui-Trong, 2000)

Figure 4.2   The urban policy neighbourhoods per department (1999) (*Source: Délégation Interministérielle à la Ville* (DIV))

geography' in that the defined criteria were used to measure the 'gap' between urban policy neighbourhoods, national means and mean values for communes and agglomerations of which they were a part. What was measured, therefore, was the concentration of long-term unemployed, young people under 25 and foreigners in relation to mean values at the commune, agglomeration and national levels. The spaces of urban policy were thus conceptualized not as separate from, but in relation to, wider geographies.

The neighbourhoods in question were the ones included in urban policy for the Xth Plan (1989–93), which were selected by local collectivities (communes and regions) and the state. The list consisted of 546 neighbourhoods, which were also the ones initially used by the RG. Table 4.1 presents some of the characteristics of these neighbourhoods.[21]

In 1990, there were 3 million inhabitants living in the priority neighbourhoods of urban policy (5.3% of the total population of metropolitan France). Most of these neighbourhoods were located at the peripheral areas of cities, and were often further separated from the city by power lines, freeways and railways. Around seven neighbourhoods out of ten were situated next to railways, although only four out of ten were served by a train station (Champion et al., 1993). More than half of the inhabitants of priority neighbourhoods lived in social housing (HLM), while this ratio was 20% in agglomerations to which they belonged, and 15% for the rest of France.

Table 4.1   Characteristics of the priority neighbourhoods of urban policy in comparison with their agglomerations and metropolitan France, 1990 (%)

	Priority neighbourhoods	*Agglomeration (city and its banlieues)*	*France total*
Unemployment rate	18.9	11.5	10.8
Unemployment rate/15–24	28.5	20.7	19.9
Unemployment rate/foreigners	26.2	19.8	18.8
Proportion of young people/15–24	18.2	15.8	15.0
Proportion of foreigners	18.6	9.0	6.3
Proportion of people with no high-school diploma (excluding students)	39.3	26.8	29.1
HLM inhabitants	59.0	20.3	14.6

Source: INSEE-DIV (no date).

The priority neighbourhoods of urban policy were places with concentrations of unemployed people, young people under 25 and foreigners. However, the situation varied depending on regional particularities, such as the economic history of a particular region. Although the urban policy neighbourhoods were marked by high levels of unemployment – almost twice the national mean – there were nevertheless neighbourhoods with equally high levels of unemployment that were not included in urban policy (Champion and Marpsat, 1996).

The proportion of young people aged between 15 and 24 was slightly higher in the priority neighbourhoods compared to the agglomerations to which they belonged. What was particularly high in the priority neighbourhoods was youth unemployment, although it must be noted that the level of youth unemployment is rather high in France, where one fifth of the young population was unemployed in 1990.

The proportion of foreigners living in the priority neighbourhoods was relatively high (more than twice compared to the agglomeration, and almost thrice compared to the national mean). Foreigners from North Africa were over-represented in the priority neighbourhoods of urban policy; in 1990, more than half the foreigners living in these neighbourhoods were of North African origin (Champion et al., 1993).

Another distinguishing feature of the priority neighbourhoods concerned the level of education. Four out of ten people did not have a high school diploma in the urban policy neighbourhoods.

This summary gives some insight into the characteristics of the urban policy neighbourhoods (and of the RG, as well) in 1990. We should, however, keep in mind that despite common characteristics (peripheral locations, concentrations of unemployed people, youth and foreigners, school failure, etc.), there were also variations among the urban policy neighbourhoods, as the first statistics about them showed (Champion and Marpsat, 1996; Champion et al., 1993). As we will see in the next chapter, all of these neighbourhoods would remain in the ever-expanding list of urban policy, but their spatial conceptualizations and discursive articulations would change remarkably.

Although these transformations of urban policy in the early 1990s provided more information about its spaces of intervention, they did something else: they defined those spaces in the first place. This relates to Rose's (1999) discussion of 'political numbers', where he explores the relationship between numbers and government. He argues that numbers make government possible 'because they help make up the object domains upon which government is required to operate' (1999: 197). In other words, they are constitutive of the spaces of government, contributing to the drawing of their boundaries and making them intelligible. Numbers, Rose (1999: 212) argues, 'do not merely inscribe a pre-existing reality. They constitute it.'

In so doing, they also problematize it: that is, define it as a problem and make it amenable to government. In other words, they help to constitute the very objects or subjects of government: 'Numbers, like other "inscription devices", actually constitute the domains they appear to represent; they render them representable in a docile form – a form amenable to the application of calculation and deliberation' (Rose, 1999: 198). Rose's argument resonates with my approach to urban policy and its spaces of intervention. Just as political numbers do not simply represent (given) domains of government but constitute them as such in the first place, the spaces of urban policy, as I have argued in Chapter 2, are not given but constituted through various practices of articulation (of which statistics is one).

Paradoxically, the recourse to numbers took place in a context where the question of *banlieues* became more politically charged, thus establishing what Rose (1999: 207) called a 'domain of objectivity' in the midst of heated debates around the issue of *banlieues*, ranging from 'integration' to terrorism. Urban policy moved away from local specificities and militant forms of intervention towards quantification, standardization and bureaucratic forms of intervention. Thus, new tools were put in place to address the question of *banlieues* in a more precise and 'objective' manner. But these new tools were also new *sensible evidences* that helped constitute the *banlieues* as objects of thought and intervention in a certain way. The statistics and mappings of INSEE, the so-called 'Bui-Trong scale' that classified 'urban violence' to create 'a cartography of sensitive neighbourhoods', and the new statistics on 'urban violence'[22] – a new term entering the agenda – were all part of that system of sensible evidences that I call 'the police'. On the one hand, therefore, new ways of thinking and talking about the *banlieues* were encouraged. On the other, there was a search for a 'better' police through the use of new tools and techniques aimed at precision.

Conclusions: Looking for a 'Better' Police . . .

Following the incidents of Vaulx-en-Velin in 1990, Prime Minister Rocard had charged the two presidents of the National Council of Cities (CNV) with a mission to visit the commune and prepare a report on the situation. The report was prepared shortly after, and it stated that the disturbances were not to be seen as a failure of urban policy, despite the fact that Vaulx-en-Velin had been part of urban policy since 1984. Urban policy had still to be pursued, the report maintained, but its conception and objectives had to be reconsidered (CNV, 1990b).

Urban policy remained a priority for the government, which showed its commitment to the issue with the creation of the Ministry for the City and the passing of new laws in the early 1990s. A remarkable feature of this period was the changing tone of urban policy and the engagement of the repressive machinery of the state with the urban question, or, better yet, with the question of *banlieues*. The initial reports on the orientation of urban policy were aimed at generating new ways of thinking about the city. Their vision of urban policy was one of political experimentation; they had hope more in militants working in the neighbourhoods than in procedural approaches. In the 1980s, social development of neighbourhoods was above all the work of these militants (see Donzelot and Estèbe's 1993 report). 'So there was really this desire to do something . . . for and with the people. With, because we were leaning on them, a network of associations' (interview, Pierre Saragoussi).

In the early 1990s, however, this approach gave way to a more technocratic and administrative one; the 'privileged interlocutor' of urban policy was now the parliament, and no longer locally elected officials (Donzelot and Estèbe's 1993 report: 101). As the members of the original DSQ programme of the early 1980s maintain, with the reforms of the early 1990s, the nature of urban policy changed; 'it became an administration' (interview, Sylvie Harburger). With the bureaucratization of urban policy and the expansion of its geography, the initial ideas of experimentation and local specificities disappeared. 'And then there is this other . . . path that was chosen [in the early 1980s], like, carrying on always with, with, specificity, [. . .] and that was lost, and we created discursive places ['*lieux de discours*']' (interview, Pierre Saragoussi).[23]

In terms of spatial conceptualization, the urban policy neighbourhoods were constituted as more precisely defined objects of intervention, with clear boundaries and corresponding statistical information. This emphasis on precision and statistical manipulation, however, was not premised on a conception of the spaces of urban policy as self-contained. Although their boundaries were now more clearly defined, they were nevertheless conceptualized in relation to wider geographies rather than as separate entities in themselves, thus transforming the 'local geography' of the 1980s into a 'relative geography'. As we will see in the next chapter, there will be another shift in the spatial conceptualization of the spaces of urban policy, transforming the relative geography of the early 1990s into a 'statist geography'. This shift will also correspond to two other shifts from the mid-1990s onwards: first, a shift from more comprehensive and potentially redistributive policies to initiatives 'confined' to the designated areas (in the form of tax concession zones, for example); and, second, a shift towards more repressive and penal forms of state intervention, the effects of which will be more pronounced in the urban policy neighbourhoods.

Before taking on this repressive orientation, however, urban policy in the early 1990s was a search for a 'better' police. Institutions of urban policy, the problem they were to tackle ('exclusion'), and the spaces they were to intervene upon were all constituted almost simultaneously in a couple of years. Equipped with new institutions, a problem to address, rationally and more accurately defined spaces of intervention, urban policy sought to operate in a more precise and efficient manner, but its spatial premises and the spatial order it consolidated were never questioned. The police order was simply expanded with more precise delineations and a new name for 'the problem' – exclusion. Thus, the inclusion of neighbour-hoods in this police order – which had proven far from temporary by the 1990s – became the very sign of their exclusion, even deviance. The remark of the mayor of Vaulx-en-Velin – made in 2002, almost two decades after having first been included in urban policy – is emblematic in this sense: 'My objective is that the city acquire one day the *status of normality*' (inter-view, Maurice Charrier).

This police order also provided the spatial basis for the engagement of the Ministry of Justice and the French Intelligence Service, which led to the production of new state statements around the *banlieues*. In a context of growing anxieties about Islam, terrorism and the formation of 'commu-nities' and 'ghettos' at the peripheral areas of cities, the urban policy neighbourhoods came to be articulated increasingly as 'menace' – especially following the engagement of the RG with the question of *banlieues*, introducing such new notions as 'urban violence', 'urban guerrilla' and 'sensitive neighbourhoods'. 'Justice' was mobilized to respond to 'urban problems'.

But other notions of justice – not merely in its penal form – were also being mobilized in the early 1990s. I have argued, in Part I, that urban policy and other state's statements are place-making practices, but not the only ones. There is, in this sense, a contestation for space between state's statements and other voices. The *Mouvement de l'Immigration et des Banlieues* (Immigration and *Banlieues* Movement, MIB) is one of the alter-native voices on *banlieues*. Created in 1995 by mainly second-generation immigrants from the *banlieues* of the Paris region, the MIB is one of the most prominent mobilizations in the *banlieues*, with an aim to federate local associations in a larger movement. As one of its founding members put it:

> Originally, the objectives and claims of the movement, you can see them on the poster downstairs, 'Justice in *Banlieues*', a programme which raises issues such as housing, access to school, *double peine*,[24] crime, and discrimination in general. (Interview, Guy, MIB)

The starting point was the identification of these 'various forms of injustice' (interview, Farid, MIB), with the objective of creating a movement in the *banlieues* which would become 'an important political force, existing as a sort of counter-power' (interview, Lahlou, MIB). In fact, what would then become the MIB – the 'Resistance of *Banlieues*' – had started around the same time as the engagement of the repressive machinery of the state with the question of *banlieues* following the revolts of the early 1990s, which had a different significance for the *banlieue* youth:[25]

> When a young person is killed in a *cité*, for the revolts of the 1990s were above all about police violence, it would be far more worrying if nothing happened. [. . .] Therefore, I always thought that it was rather a phenomenon of resistance that exposed the limits of political discourse. [. . .] Revolt is the only remaining possibility to say at a given moment, that's enough, we're capable, let's go. [. . .] The period of 'Resistance of *Banlieues*' was so clear in that way, revolts were taking place daily, that was between the late 1980s and early 1990s. (Interview, Guy, MIB)

As we will see in more detail in Chapter 7, most of the *banlieue* revolts are triggered by the killing of *banlieue* youth in circumstances where the police are implicated. As we have seen above, the state's statements of the early 1990s highlighted mainly a criminal dimension in revolts, and articulated the *banlieues* with new notions, such as 'urban violence', 'urban guerrilla' and 'sensitive neighbourhoods'. 'Resistance of *Banlieues*', on the other hand, was created, as one of its founding members put it, 'to make people hear another voice, and to make claims about what the people needed in these so-called sensitive neighbourhoods, about which the only sensitive thing is the misery and precarious situation of people' (interview, Nordine, MIB).

Such voices, however, have not been very effective in changing the discursive terms of *banlieues*, and we will see some of the reasons why in Chapter 6. In addition to financial and organizational difficulties, such alternative voices originating in the *banlieues* have increasingly been undermined by accusations of 'communitarianism', which has become a common practice from the early 1990s onwards with the rise of republican nationalism.

... a 'Republican' One

The early 1990s were also marked by the consolidation of what Blatt (1997) called 'the republican consensus', with references to so-called republican 'values', 'principles' and 'tradition', and the development of a so-called 'republican model of integration'. Despite the rise of republican nationalism in the early 1990s with considerable hostility towards different, notably non-European, cultures, the focus remained on the National Front (FN) and its leader, Jean-Marie Le Pen, which became the major target of anti-racist movements.[26] But Le Pen was not the only politician deploying racist discourses. In 1991, Jacques Chirac, then mayor of Paris, expressed his concern about immigration in quite explicit terms, stating that 'the threshold of tolerance' had been surpassed, and that there was an 'overdose' of immigrants:

> The worker who [. . .] works with his wife to earn around 15,000 francs and who sees, on the same floor of his HLM, a packed family with the father, his three or four wives and some twenty kids, who makes 50,000 francs through welfare payments without, naturally, working . . . if you add to this the noise and the smell, the French worker goes crazy. Saying this does not mean being racist. (*Le Monde*, 21 June 1991: 40)

Chirac did not regret having said this. In an interview the following day, he claimed that this was what was heard 'in these neighbourhoods', and stated, by using precisely the same reasoning Le Pen uses to substantiate his own racist remarks, that all he did was to 'say loudly what many think': 'I am not suspected of sympathizing with Le Pen, I do not see why he would have the monopoly over highlighting the real problems.' Le Pen, on the other hand, was 'surprised to see that more and more people subscribe to the analysis and the proposals that the National Front has been making for 10 years, while still demonizing it' (*Le Monde*, 21 June 1991: 40).

Chirac's words, coming from the mouth of a former Prime Minister, the mayor of Paris and the president of the opposition party RPR at the time, were not without political weight. However, Le Pen and FN remained as the symbols of racism in French political life. Chirac, on the other hand, was elected President of the Republic four years later, in 1995. Ironically, when he was faced with Le Pen during the second round of 2002 presidential elections, he presented himself as defending 'republican values' and the 'republican tradition of tolerance and democracy', and was eventually elected President of the Republic for the second time.

The republican consensus was centred on the alleged 'threat' of the formation of 'communities' deemed incompatible with the principles of the French republic, which was spatially reified in the image of the *banlieue*. In the early years of urban policy, the presence of different cultures was seen as a condition of urban life, and the diversity engendered by this presence was celebrated. There were even occasional references to 'right to difference', until the notion's appropriation by the extreme right,[27] and its official denunciation by the HCI's second annual report in 1992. The report stated that '[n]otions of a "multicultural society" and the "right to difference" [were] unacceptably ambiguous', and that 'it would be wrong to let anyone think that different cultures can be allowed to become fully developed in France' (HCI, 1992; cited in Hargreaves, 1997: 184). This was a clear message, but a more brutal one was delivered by Charles Pasqua, a couple of days before the first round of 1993 legislative elections. He declared that it should not be

> accepted that certain communities be constituted, that they refuse our culture, and try to impose theirs on us, their customs and habits. . . . If France does not suit them, all they have to do is go home and bugger off [*foutre le camp*]. . . . We must not tolerate in our society the evolution towards a multi-ethnic, multi-racial and multi-cultural society. Multi-ethnic and multi-racial, yes, but not multi-cultural. Those who want to live on the national territory must become French and assimilate our culture, we don't have to put up with the others. We must both show openness and welcome all those who want to become our brothers, and reject those who accept neither our culture nor our civilization, and who, in the long term, constitute the seed of serious ethnic or racial difficulties. (*Le Monde*, 21–2 March 1993: 11)

This was a far cry from Dubedout's formulation for the social development of neighbourhoods (DSQ) programme:

> The cultural element is essential in this respect. To foster the necessary conditions for the emergence of 'other' cultures, considered today as minor, even ignored or despised, to provide social groups that do not identify with 'high' culture with the means to assert their cultural identity and way of life – such is the ambitious aim towards which urban societies should strive to take up the challenge of their multiracial condition and their own future. (Dubedout Report, 1983: 60)

It was in this context of a renewed enthusiasm for the 'republican tradition' that a centre-right government came to office in 1993, winning a sweeping majority in the parliamentary elections of 21 and 28 March, with Edouard Balladur as Prime Minister, Charles Pasqua as the Minister of the Interior,[28] and Simone Veil as Minister for Social Affairs, Health and Urban Affairs, taking the former Ministry for the City under her ministry.

5

From 'Neighbourhoods in Danger' to 'Dangerous Neighbourhoods': The Repressive Turn in Urban Policy

The first parliamentary debate of the new government was devoted to cities and *banlieues*, which was rather uncommon for a right-wing government. The *banlieues* were immediately on the first page of the newspaper *Le Figaro* on 27 April 1993(a), with the following title: 'As the debate on the *banlieues* opens at the Assembly, Pasqua wants to re-establish identity checks', although the article was not really about *banlieues*, but about two new laws on security and immigration that Pasqua wanted to present at the Ministry Council.

There was another article, again on the front page, by Alain Peyrefitte, former Minister of Justice and the author of the 1977 report on violence, which we have discussed in Chapter 3. Linking unemployment, insecurity, immigration and the 'malaise of *banlieues*', Peyrefitte accused the left of a soft, even naïve, attitude, and pointed to a widespread 'fear' of HLM *cités* and *banlieues*, which he qualified as 'outlaw areas' (*zones de non-droit*)'. The solution required 'firmness in the respect of republican principles'.

At the National Assembly, the issue of the 'problems of cities and *banlieues*' was presented as one of the priorities of the new government, which, again, was pretty uncommon for a right-wing administration. A so-called 'Marshall Plan' was declared for *banlieues*. Prime Minister Balladur had already identified the causes of the 'malaise of *banlieues*' as overpopulation, degraded housing, school failure, unemployment, insecurity and 'communitarian co-existence' (*Le Figaro*, 27 April 1993b: 1). During the parliamentary debate, he argued that 'the spirit of the republic' and 'the authority of the state' were 'challenged' ('*mise en cause*'). He affirmed the necessity to reaffirm the authority of the state, particularly in the so-called 'outlaw areas'. The three priorities envisaged by his party, the RPR, were 'authority, activity, and identity'. Authority meant the 'return of the state in the *cités*'; activity referred to tax benefits for firms locating in the urban policy neighbourhoods; and identity meant integration by 'giving renewed pleasure at being French to those who become so by chance' (*Le Figaro*, 28 April 1993: 6). This last priority reflected a series of new laws in preparation, which passed three months later. Known as the 'Pasqua–Méhaignerie Laws',[1] they restricted entry and residence rights of foreigners with a stated aim of 'zero immigration', and facilitated easier identity checks by the police. But more importantly perhaps, they introduced a revision of the French Nationality Code. No legislative reform of the code had passed in the 1980s, despite the (failed) attempts of the Chirac government of 1986–8. This 1993 nationality reform defined children born in France of immigrant parents as foreigners, whereas before the reform they would almost automatically have acquired French citizenship upon birth. The reform required immigrant youth to 'manifest their wish' to become French citizens between the ages of 16 and 21 (see Feldblum, 1999, for a detailed account of the reform).

The discourse of the new government diverged remarkably from the earlier governments' discourse on the *banlieues*. Linking the *banlieues* with issues such as 'the spirit of the republic', 'the authority of the state' and 'French identity' was something virtually absent from the state's statements in the 1980s. Whereas earlier policies were more concerned with alleviating inequalities rising from the economic crisis, the new government proposed a shift in the legitimation of state intervention. With the introduction of new notions such as 'outlaw areas' and the 'malaise of *banlieues*', *banlieues* evoked an image of imminent 'menace', and were no longer spaces where one could identify the signs or symptoms of larger problems. The 'threat' to the integrity of the republic had found its spatial form: 'the *banlieue*'.

Encore! The Ghost Haunting the French Republic

> [U]rban policy definitely contributed to limit effects of urban segregation. By improving public service in the neighbourhoods, it prevented a dangerous trend [*dérive dangereuse*], like the one experienced by Anglo-Saxon countries.
>
> Geindre Reprort, 1993: 42

The consolidation of the republican consensus was easily discernible in the next urban policy report. The Geindre Report of 1993 was part of the work of commissions working for the preparation of the XIth Plan (1994–8). The report was critical of the ambiguity of urban policy and its lack of clear objectives. The report's major proposition concerned the focus of urban policy; it had to be 'fighting against exclusion'. Like many of the previous reports, the Geindre Report maintained that the objective of participation of inhabitants had not been achieved. Although it was recognized that almost half of the inhabitants in the urban policy neighbourhoods were excluded from formal participation through voting (either because of age or nationality), the report did not opt for voting rights for foreigners. Overall, it was stated that urban policy had been helpful in avoiding 'dangerous evolutions, the formation of ghettos or outbursts of violence' (Geindre Report, 1993: 236). Since Vaulx-en-Velin, however, 14 large-scale revolts had taken place in social housing neighbourhoods, and another 12 a year after the publication of the report. All but two of the areas where incidents occurred were urban policy neighbourhoods, and all except one were located in the *banlieues*.

What really distinguished the Geindre Report, published a year after the Los Angeles riots, was its emphasis on the republic, its values and authority, and the 'threat' posed by the formation of 'communities'. The problem had changed direction from the neighbourhoods themselves to the republic; once the neighbourhoods were 'at risk', now it was 'the republic':

> Some neighbourhoods are escaping *republican law*. Problems of ethnic co-existence should not be evaded. [. . .] Immigrant workers of the *'trente glorieuses'* had the opportunity to integrate into French society, but what are the opportunities for unemployed immigrants or children of immigrants? In ravaged neighbourhoods, isn't there already a quest for 'community' identity, to the image of what happens in the United States or Great Britain, and isn't there a risk that secular principles and *republican values* be challenged? (1993: 8; emphasis added)

This was a remarkable change because 'the republic' was virtually absent from the previous statements of the state in the domain of urban policy, except for a brief reference in the 1991 Delarue Report. The Geindre Report was the first urban policy document in which the neighbourhoods were explicitly seen as potential 'threats' to the republic and its values. Thus conceiving the problem, or the 'risk', the Geindre Report emphasized the necessity of a strong territorial presence of the state through its *'politiques régaliennes'* and *'services régaliens'*. The adjective *'régalien(ne)'* is significant. The *services régaliens* of the state designates state functions (such as administration and finance) and state authority (such as the police and the justice system). It was the latter that was advocated by the report:

> The reassertion of the rule of law [*état de droit*] in the difficult neighbourhoods [*quartiers difficiles*] implies for the state to take steps to both make everyone respect the law and give everyone the possibility to have his/her rights

respected. [. . .] It is therefore necessary to ensure a strong territorial presence of state services [*services régaliens*]. The aim is to make the rule of law visible mainly by locating police stations and justice representatives in the toughest neighbourhoods. Ensuring that the law be respected implies an increased presence of police forces in the field and a quick and adapted judicial response. (1993: 81)

The Geindre Report reflected the changing tone of state's statements in the domain of urban policy with its emphasis on 'the republic' and the authority of the state. This trend intensified following the change of government. In May 1995, Jacques Chirac was elected President of the Republic following a campaign organized around the theme of 'social fracture'. Alain Juppé replaced Balladur as the Prime Minister, and Jean-Louis Debré replaced Pasqua as the Minister of the Interior. First put into circulation in the early 1990s, the notion of 'urban violence' immediately became a priority for the new government. In a circular addressed to the prefects, Debré qualified the 'control of urban violence' as 'a priority action'. The phenomena of urban violence, he stated, 'must be taken into account rapidly and efficiently, failing which they will become a major factor of trouble in the *banlieues* likely to jeopardize the balance of our society'.[2]

While the new Ministry of the Interior was occupied with responding to 'urban violence' with a focus on the *banlieues*, urban policy-makers were already busy with the delimitation of priority neighbourhoods for the XIth Plan (1994–8). Francis Idrac, a prefect, was now the head of DIV. In a circular addressed to prefects, he asked them to indicate the 'precise location of neighbourhoods' on maps, and to establish a hierarchy of the areas delimited. The attempt was to establish the 'geography of the priority neighbourhoods of urban policy'. The criteria were set as the following:

-   gap between the situation of the neighbourhood and the rest of the urban areas with respect to the proportion of young people under 20, unemployment rates, proportion of immigrant population;
-   mono-functionality of the area with an imbalance between residents (*grands ensembles* and degraded housing) and economic activity and employment;
-   isolation [*enclavement*] of the area with respect to the rest of the city and in terms of access to public services.[3]

The following year, a report on 'urban integration' was prepared by Idrac and Jean-Pierre Duport, another prefect. Their diagnosis was that the 'social cohesion' of French cities was 'threatened' by segregation, unemployment and urban violence. The envisioned programme for urban integration was basically concerned with 'neighbourhoods in difficulty' and '*cités*', which were seen to be turning into 'ghettos': '[T]he phenomenon of transformation of certain marginalised urban areas into ghettos continues and becomes more pronounced' (Programme National d'Intégration Urbaine, 1995: 1). The notion of ghetto was now part of the official discourse, although other statements of the state contradicted this assertion. The results of the first statistical profiling of the urban policy neighbourhoods showed that they were neither ethnically homogeneous nor large enough to function as self-contained segregated areas (see, for example, Champion and Marpsat, 1996; Champion et al., 1993). The image of the ghetto, however, was a powerful one to summon the ghosts haunting the French republic: 'Life in the *cités* tends to breed exclusion and to express itself by behaviour beyond the cultural and social realm of the republican model' (Programme National d'Intégration Urbaine, 1995: 1).

Based on this articulation of the danger menacing the social cohesion of French cities, Duport and Idrac defined the four major objectives of the National Programme for Urban Integration as the following: (i) developing activity and employment in neighbourhoods; (ii) diversification and restructuration of neighbourhoods; (iii) development of associations; and (iv) restoring the presence of the state and of public services. This programme would form the basis of a new law the following year, the *Pacte de Relance pour la Ville* – presented by President Chirac as a 'Marshall Plan for the *banlieues*'.[4]

Pacte de Relance: Old Ghost, New Spaces

Unlike the United States, we cannot accept that portions of our territory be definitely excluded, abandoned to others, and drift into violence and misery under the disillusioned or cowardly eyes of public powers.

> Gérard Larcher, spokesperson for the
> special commission on *Pacte de Relance*

Indeed, [. . .] this *Pacte de Relance* is in a way the last chance to avoid what occurs in other countries, ghettos and conflicts between communities.

> Jean-Pierre Fourcade, president of the
> special commission on *Pacte de Relance*

André Vezinhet (Senator, PS): But how shall we avoid troubles similar to that experienced by England, where the enterprise zones policy had to be given up because of its inefficiency?
Jean-Claude Gaudin (Minister, UDF): That's in England![5]

The spatial focus of the *Pacte de Relance* did not differ from the previous programmes. Its aim was presented by Prime Minister Juppé as the better integration of the so-called 'neighbourhoods in difficulty'. It was conceived, in the words of Jean-Claude Gaudin, the Minister for Territorial Planning, City and Integration, to 'respond to the problems of *banlieues*'. The aim, he stated, was to fight against 'territorial fracture' – the spatial ramification of Chirac's electoral campaign theme, 'social fracture' – which drifted parts of the population away from 'the republic' and 'its values'. The *Pacte de Relance*, he affirmed, was a 'republican pact'. Everything in the 1990s, it seemed, was 'republican', except the *banlieues*, which were deemed to somehow defy the republic's values and principles.[6]

The major features of the *Pacte de Relance* were defined by Idrac, head of the DIV. He stated that the *Pacte* was aimed at reducing 'social fracture' as its primary objective. This fracture consisted of four main lines. The first was housing, notably social housing. In addition to physical degradation, the problem identified by Idrac was the populations living in the social housing estates. Social housing estates were mainly inhabited by poor segments of the population. But more importantly, the concentration of immigrants and people from immigrant backgrounds in certain areas, Idrac held, led to a process of 'ghettoization' (1996: 20). The second line of fracture was unemployment, and the third the inability of young people to integrate themselves economically into society. The fourth was 'the phenomena of urban violence', which, in the words of Idrac, was a 'characteristic element of contemporary *cités*' (1996: 20).[7] And these lines of fracture were produced in a social context in which 'traditional bearings' were destabilized. Parental authority, for example, no longer played its role as the 'traditional regulator'; the 'image of the father' was devalorized with unemployment; the credibility of political organizations and syndicates was lost; and religious values were largely absent, although there was a revival of 'certain forms of fundamentalism' (1996: 21). Such were the 'elements of diagnosis' provided by the head of the DIV.

In order to respond to these problems, Idrac identified four major axes for the *Pacte de Relance*. The first and foremost of these was the 'revival of activity in the neighbourhoods', which basically consisted of tax concessions and subsidies to business locating in the designated areas, and the creation of jobs for young people aged between 18 and 25. The second major axis of the programme was 'security and public peace'. Idrac stated that the novelty of the new programme was that policies of prevention – the focus of previous urban policies – would be complemented by policies of 'security, public tranquillity and public peace', with improved police presence in the *banlieues* and a more effective judicial response (1996: 30). To this end, 4,000 new police officers would be assigned to communes with 'difficult neighbourhoods'. The other two major axes of the programme were education, which implied a coordinated action in the priority neighbourhoods with the Ministry of Education, and urban integration, which was aimed at achieving 'social mixity' in these neighbourhoods, which were seen to be in a process of social and 'ethnic' segregation. The encouraging local political formations and the democratization of the management of the city, which were at the heart of Dubedout's original conceptualization of urban policy, did not figure among the major axes of the programme. Dubedout, a mayor involved in the *autogestion* movement, was concerned with the inhabitants and their democratic aspirations. Idrac, a prefect, was concerned with the order and cohesion of the republic. 'The *Pacte de Relance pour la Ville*', he concluded, 'is at the service of the Republican Pact, which is the basis of the social cohesion of the Nation' (1996: 34).

The *Pacte de Relance* was conceived with these concerns, which highlighted 'the problem of *banlieues*' more in terms of 'ethnic' origins and public order. The main tenet of the programme was the 'differential' treatment of the urban policy neighbourhoods by providing financial advantages for businesses in these areas. The name given to this differential treatment, however, could not be 'affirmative action' for it connoted too strongly the wrong-headed approach of the so-called 'Anglo-Saxon model'. Thus an oxymoron was produced: 'positive discrimination', the French version of affirmative action.

'They are Already Stigmatized': Affirmative Action à la française

> *Guy Fischer (Senator, PC):* The historic model of positive discrimination is American: it is based on an objective identification of disadvantaged minorities – Blacks, Chicanos, women . . .
>
> *Gérard Larcher (spokesperson of the Pacte de Relance, RPR):* So women come after Chicanos? *(laughter and shouts)*
>
> *Guy Fischer:* . . . whose social handicaps it aims to compensate by establishing quotas in schools, universities and public employment.
>
> *Jean-Claude Gaudin (Minister, UDF):* No quotas! *(smiles)*
>
> *Gérard Larcher:* Quotas are an Indian tribe, aren't they? *(more smiles)*[8]

Trying to foster economic activity and to increase employment through tax concessions was not a novelty brought about by the *Pacte de Relance*. Attempts had already been made in 1987 by the creation of *zones d'entreprises*, inspired by Margaret Thatcher's enterprise zones, by the then Minister of Industry, Alain Madelin, in the framework of the admittedly neoliberal programme of the 1986–8 Chirac government. The LOV of 1991 had also envisaged such a measure for businesses locating in the designated neighbourhoods, although it was never widely applied (Estèbe, 2001). Finally, Charles Pasqua had created the *Zones Urbaines Sensibles*

(ZUSs) and *Zones de Redynamisation Urbaine* (ZRUs) with the so-called 'Pasqua Law' of 4 February 1995.[9] With this law, tax concessions were introduced for new business locating in the designated areas. What the *Pacte de Relance* did was to appropriate this measure by creating a third category, the *Zones Franches Urbaines* (ZFUs), and extend tax concessions to already existing firms as well.

The law was passed on 14 November 1996 with a stated aim to 'fight against the phenomena of exclusion in urban space, and further the professional, social, and cultural insertion of populations living in *grands ensembles* or degraded residential neighbourhoods'. The ZUSs were defined as areas 'characterized by the presence of *grands ensembles* or areas of degraded housing and by a major imbalance between housing and employment'. They comprised the ZRUs and the ZFUs. The ZRUs were

> those of the *Zones Urbaines Sensibles* [ZUSs] defined above that are confronted with particular difficulties to be appreciated according to their location in the built areas, the economic and commercial characteristics and according to a synthetic index. This index is established, according to conditions laid out in a decree, taking into account the number of inhabitants of the area, the unemployment rates, the proportion of young people under 25 years old, the proportion of school drop-outs, and the fiscal potential of the relevant commune.

The ZFUs were defined using the same criteria as the ZRUs, but included 'particularly deprived' areas with more than 10,000 inhabitants.[10] With the *Pacte de Relance*, 700 sites were designated as ZUS (containing some 1,300 neighbourhoods), of which 350 were ZRUs, and 38 were ZFUs. The neighbourhoods that were already included in urban policy remained, while new ones were added. With the addition of new neighbourhoods, the population living in the urban policy neighbourhoods reached 4.7 million – around 1 inhabitant in 12 for the total population of metropolitan France.

The special treatment of these areas was legitimized through the introduction of a new notion. Appropriating the notion of affirmative action to the image of the republic, the *Pacte de Relance* officially introduced affirmative action *à la française*: 'territorial positive discrimination'. 'Positive discrimination' based on culture, ethnicity or religion, as we have seen in Chapter 1, is contrary to the republican principles and is unconstitutional. Given this, it could only be *territorial* positive discrimination. This is what separates the republic from the so-called 'Anglo-Saxon' model, which is to be avoided at all costs. As Idrac, one of the conceivers of the *Pacte de Relance*, (wrongly) explained: 'The Anglo-Saxon discrimination is embedded in a communitarian logic that recognizes ethnic communities as such.

Such an approach would sharply be contrary to the constitutional and republican principles of France' (1996: 25).

Designating spaces, on the other hand, was not contrary to the principles of the republic, since it did not imply the official recognition of any particular groups based on culture, ethnicity or religion. Space, in the statist geography of the *Pacte de Relance*, seemed both homogeneous and homogenizing. Such an exclusive focus on designated spaces, however, carried the risk of (further) stigmatization. Since the Dubedout Report, almost all urban policy reports had been cautious about the perils of such an approach, the stigmatizing effects of which were largely recognized. The *Pacte de Relance* not only continued this approach, but also put the urban policy neighbourhoods on a scale as absolute spaces of exclusion, albeit to varying degrees. Idrac recognized the risk of stigmatizing certain neighbourhoods, but justified the approach by stating that it was obvious to see that they were 'already stigmatized' (1996: 27).

In spatial terms, the basis of urban policy did not change; it still defined and intervened in a 'geography of priority neighbourhoods' – what I have referred to as 'the police'. But the *Pacte de Relance* brought about a major change in the conceptualization of the spaces of urban policy, and turned the previous 'relative geography' of priority neighbourhoods into a 'statist geography', defined in absolute and hierarchical terms through centrally determined criteria. This change again required a collaboration with the INSEE. For the most part, the *Pacte de Relance* retained areas that were already included in DSQ and City Contracts programmes. The novelty of the programme was what it did with these areas. A formula called 'Synthetic Index of Exclusion' (*Indice Synthétique d'Exclusion*, ISE) was devised to categorize neighbourhoods and assign them to their proper places on a scale of exclusion. In the 'relative geography' of the 1990s, the attempt was to measure the gap between the designated neighbourhoods and their surrounding areas. The designated neighbourhoods were seen as 'neighbourhoods at risk', depending on the size of the gap, and the attempt was to discern symptoms of larger problems and populations at risk (Estèbe, 2001). This time, however, the attempt was to measure how badly these neighbourhoods were 'excluded'. As Béhar (1998: 3) argued:

> For the City Contracts, the criteria used to define the priority geography (unemployment rates, proportion of young people, of foreigners, of school drop-outs) could identify the signs of a problem, the roots of which are to be found elsewhere. For the *Pacte de Relance*, they are the very roots of the problem – an excessive concentration – to be reduced.

The calculation of the degree of exclusion was based on the ISE, which took into consideration the level of long-term unemployment,[11] proportion of young people and proportion of people without a high-school diploma, which replaced the proportion of foreigners used in the previous calculations.[12] What the *Pacte de Relance* did was to constitute the urban policy neighbourhoods as spatial categories of exclusion, and to create a hierarchical geography of priority neighbourhoods, some of which more excluded (i.e. ZFUs) than the others. It was premised upon a view of space geometrically divisible into discrete parts, which could then be categorized and placed on a scale. The remarks of Jean Pirot, the project manager in Vaulx-en-Velin (which became a ZFU with this programme), are exemplary here:

> I remember the *Zones Franches* [ZFUs], I was working here, wasn't I. You'd see, it's a pity I no longer have the map, the limits of the *Zones Franches*. The *Zone Franche*, it was unbelievable, it was part of the city, with one side of certain streets included, the other side excluded! So a very constrained, a very limited geography. (Interview, Jean Pirot)

The decree that defined Vaulx-en-Velin's ZFUs gives a sense of what Pirot is referring to:[13]

- intersection between the bypass and 8-Mai-1945 avenue (parcel section AW n 308) until A42 motorway;
- A42 motorway until Louis-et-Marie-Louise-Baumer street;
- Louis-et-Marie-Louise-Baumer street until Balmont path;
- Balmont path until the western limit of the parcel section AV n 2;
- western limit of the parcel section AV n 2 until Pierre-Cot street;
- Pierre-Cot street until Henri-Barbusse avenue;
- Henri-Barbusse avenue until the northern limit of the parcel section AX n 293;
- northern limits of the parcels section AX nos 293 and 294 until Georges-Rouge avenue . . . [the list continues for another 46 lines]

In an interview before the passing of the law, the mayor of Vaulx-en-Velin, Maurice Charrier, criticized the *Pacte de Relance* as follows: 'Most of the proposed measures are but re-formulations of old methods that have long shown their limits because they lead nowhere: managing the problems of the ghetto, within the ghetto, by the ghetto' (interview in *Libération*, 19 January 1996: 6).

Although Vaulx-en-Velin is hardly a ghetto, Charrier's remarks are telling; the new programme closed the neighbourhoods upon themselves. The *Pacte de Relance* was arguably the closest French urban policy got to a neoliberal approach, with a shift in focus from solidarity between communes to economic success within strictly defined spaces of intervention.

It was conceived mainly in economic terms, with an exclusive focus on delimited neighbourhoods as neatly defined, absolute and calculable spaces of intervention. This spatial conceptualization meant a farewell to redistributive policies conceived to foster collaboration between communes within the larger city-region. Solidarity between communes in terms of finance and provision of social housing, which was an important – if unrealized – feature of the LOV, had disappeared. The attempts to open up the spatial focus of urban policy by considering the neighbourhoods in relation to their larger city-regions came to a halt. For the *Pacte de Relance*, the designated spaces of urban policy existed in and of themselves as neatly delimited areas that supposedly contained both the problem and its solution. In the process, the earlier ideals about the political implication of inhabitants in the production of their lived spaces vanished. The inhabitants and local specificities turned into internally homogeneous spatial categories, which led Béhar (2001) to argue that urban policy, despite all its spatial focus, was in fact an 'a-territorial' policy in that it became disconnected from local specificities.

Despite the neoliberal orientation of the *Pacte de Relance*, however, it did include a government-sponsored work programme. Called '*emplois de ville*', this programme was aimed at creating jobs for young people aged between 18 and 25, with no university degree, and living in designated areas with *grands ensembles* and degraded housing, which were basically the urban policy neighbourhoods. The state's financial engagement, however, was degressive, which meant increasing costs for the local collectivities. The state's share in the subvention of these jobs would either be 75% for the first year with a 10% decrease in each successive year or a constant 55% for a total period of five years. The aim was to create 25,000 jobs a year, although in the first year of the programme, only half of this objective was achieved (DIV, 1999). Moreover, the programme meant increasing costs for the local collectivities as they had to participate in the subvention of jobs. The Jospin government that took office in 1997 changed the programme into a general youth employment programme for people under the age of 30, and augmented state subvention to 80% of the minimum wage plus all social security contributions. Called '*emplois jeunes*', this programme provided full-time employment for a period of five years,[14] whereas the previous programme provided either full- or part-time jobs for one year, renewable each year for a total of five years. Twenty per cent of the jobs created within the framework of this programme would be for young people either from the urban policy neighbourhoods or working for them (CIV, 1998a). However, the Raffarin government that took office in 2002 saw this programme as a 'social treatment of unemployment', modified it by allowing the contracts to end once their five-year term expired, and replaced the programme with a smaller one ('*contrat jeunes en entreprises*') which was aimed at placing unskilled youths in private-sector jobs (see Levy, 2005: 186–7).

Is 'Positive Discrimination' Negative?

A year after the *Pacte de Relance*, the left was back in power in June 1997 under the cohabitation, with Lionel Jospin (PS) as Prime Minister. It was, once again, necessary to define the orientations of urban policy. A report was demanded from a commission chaired by Jean-Pierre Sueur, mayor of Orléans. The Sueur Report, *Demain, la Ville*, was published in 1998. It is possible to discern five main features of this report. The first two involve a critique of the economic measures brought about by the *Pacte de Relance*, and, more generally, the idea of 'zoning'. The third is about the 'proper scale' of urban policy, and the last two reflect the republican consensus that seemed to have characterized urban policy discourse since the early 1990s with its emphasis on the republic and the authority of the state.

As we have seen, the *Pacte de Relance* was conceived almost exclusively in economic terms with the idea of attracting business to designated areas through tax concessions. The idea behind this measure was that jobs would eventually be created for the inhabitants of designated areas. However, the Sueur Report argued that this measure faced two problems. First, most of the firms moving to these areas were of small-scale, and did not create many jobs. Second, when larger ones moved in, they did so with their own employees rather than providing jobs for the inhabitants of designated areas. The *Pacte de Relance* required the firms to employ one fifth of their employees from the designated areas where they were located. But no effective measure was envisaged to monitor the obligation to employ locally; assessment was based on declarative data, provided by the firms themselves.

A similar observation was made in an OECD report on 'integrating distressed urban areas'. The *Pacte de Relance*, it was argued, demonstrated the potential dangers and unwanted side-effects of actions oriented on spatially delimited areas with the aim of creating jobs and economic activity. The report's findings for the ZFUs created in the framework of the *Pacte de Relance* were as follows: a relatively small number of businesses were created; the firms attracted to these areas did not necessarily have a profile of activities to contribute to economic revitalization in these areas; in some cases, available commercial real estate was occupied by industrial enterprises, making the area even less attractive as a residential district; jobs created in such areas were often transferred in, and there was no guarantee that new jobs would go to the inhabitants of designated areas; and, finally, the 'ZFU' labelling contributed to the stigmatization of these areas, drawing away the very thing that it was conceived to draw in: that is, businesses (OECD, 1998: 126).

This last point resonated with the Sueur Report's main criticism of the spatial approach of urban policy: the so-called 'positive discrimination', introduced and institutionalized by the *Pacte de Relance*, was not all that 'positive' after all. Not only did the designated areas to be positively discriminated have fewer resources in terms particularly of public services (for example, 40% of them did not have a post office), but their very spatial designation as 'zones' stigmatized these areas even further. 'No one wants to live in a "zone". "Zoning" a space is very often to contribute to its disqualification, even with the best intentions' (Sueur Report, 1998: 26).

'Zoning', however, has never been in short supply in French policies, and the 'zones' designated by policies did not always necessarily carry negative connotations: *Zones à Urbaniser en Priorité* (ZUPs) in the 1960s, replaced after 1967 by the *Zones d'Aménagement Concerté* (ZACs), the *Zones d'Education Prioritaires* (ZEPs) since 1981, the *Zones Urbaines Sensibles* (ZUSs) and *Zones de Redynamisation Urbaine* (ZRUs) since 1995, and the *Zones Franches Urbaines* (ZFUs) since 1996. Although some of the previous reports on urban policy were sensitive to the issue of an exclusive spatial delimitation and designation, they had merely pointed to the risk of stigmatization without really denouncing the approach as such. The Sueur Report, in this sense, involved a 'radical' critique of urban policy: 'Twenty years of urban policy have taught us that a neighbourhood could not be changed by enclosing it within limits, within its perimeter. With "zoning" one stigmatizes as much as one helps the "zones" concerned' (1998: 171).

The problem was not just designating 'zones', but assigning them certain qualities. As we have seen in the first chapter, although the term 'zone' carries negative connotations, the way it was first used in policies did not necessarily convey a negative image. Maybe this was an unfortunate choice of terms given the negative connotations of 'zone' in daily language. The evolution of urban policy discourse thus far suggests that it was not spatial designation as such that automatically assigned such areas a negative image, but their discursive articulations in increasingly negative terms over the years.

The Sueur Report's recommendation was that the 'proper scale' of urban policy was the whole agglomeration, and not merely the neighbourhood. What was recognized was not only the stigmatizing effects of designating neighbourhoods, but also the larger dynamics of the city-region. Although the negative effects produced by such dynamics were concentrated and more easily discernible in certain areas, urban policy had to be opened up, and comprehensive policies in other realms – not merely in the domain of urban policy – had to be devised.

Despite its critique of the spatial focus of urban policy and its call for comprehensive policies – which were fairly innovative – the Sueur Report nevertheless shared the same republican consensus that had characterized urban policy since the early 1990s. Unlike the previous reports prepared by Socialist mayors in the 1980s, the Sueur Report passed in silence over issues such as the conflictual nature of urban life, discrimination and racism. A warning as to 'communitarian sociabilities incompatible with the republican model' (1998: 157) was made, and the integration of foreigners and people from 'foreign backgrounds' was stated as a major challenge posed to '*the French Nation*' (1998: 205; emphasis in original). In terms of the authoritarian functions of the state, the report called for an increase in the number of *Maisons de Justice et du Droit* (MJDs) in order to render the law 'visible and legible on the terrain'. The rapid establishment of MJDs in the urban policy neighbourhoods was proposed, with an objective to increase their number to 200 in the following two years.[15] The report also supported a new measure called 'local security contracts', and called for its generalization in the communes concerned by urban policy. In addition, the number of police forces in 'sensitive neighbourhoods' was deemed inferior compared to the rest of the country in terms of the populations concerned, and a recommendation was made to achieve 'equality' in this sense (1998: 200).

Although the Jospin government did not opt for comprehensive sectoral policies, it nevertheless tried to open up the focus of urban policy by reintroducing the City Contracts programme, which was first introduced in 1989, and passing a new law that sought to redistribute social housing more fairly, which was a revised version of the LOV of 1991. However, its more immediate actions after taking office concerned the issue of security, with 'the police' consolidated by urban policy as the spatial basis.

Insecurity Wins the Left: The Villepinte Colloquium

In less than five months after coming to power, the Socialist government organized a colloquium with the initiative of the Minister of the Interior, Jean-Pierre Chevènement, which took place in Villepinte (Seine-Saint-Denis) on 24–5 October 1997. Called 'Safe Cities for Free Citizens', known commonly as the Villepinte colloquium, this highly publicized event marked a major turning point in the attitude of the left towards the issue of security. In his inaugural speech, Chevènement delivered a clear message: '[T]oday there are two threats the republic has to face up to: unemployment and insecurity' (Chevènement, 1997: 13). In his concluding speech, Prime Minister Jospin expressed the determination of the government to deal with the issue of insecurity, which, he stated, would be a priority issue after unemployment. 'There cannot be', he said, 'on the one hand, safe neighbourhoods and, on the other, outlaw areas [*zones de non-droit*]'. Insecurity, Jospin held, was a form of inequality, and the 'right to security' would follow from 'the republican principle of equality'. His spatial references were 'outlaw areas', 'difficult areas' and *banlieues* (Jospin, 1997).

Jospin also announced the creation of a new institution – the Interior Security Council (*Conseil de Sécurité Intérieure*), chaired by the Prime Minister – and the implementation of a new measure – Local Security Contracts (*Contrats Locaux de Sécurité*, CLSs hereafter). Three days later, the Minister of the Interior's circular on the implementation of CLSs was out. The CLS's primary objective was the 'sensitive neighbourhoods'. Elaborated conjointly with the prefect, state prosecutor and the mayor, the first step in the preparation of a CLS involved a 'local security diagnostic' based on three criteria: a report on the current situation in terms of delinquency; 'an appraisal of the feeling of security'; and assessment. The circular also stated that new recruitments would be made in the National Police in order to be deployed in 'the most sensitive areas', and more MJDs would be created in order to 'reintroduce the law in areas where it has disappeared'.[16]

The Villepinte colloquium and the prioritization of the issue of security were a major turn for the left. This partly had to do with the municipal elections of 1995, after which the left had lost many cities to the right, whose campaign was centred on the issue of insecurity. Consequently, there was a pressure on the government from local officials to address the issue, which it did, perhaps too zealously. In any case, its attitude was unprecedented for a Socialist government. Following the 1997 elections, many people expected a strong political gesture from the new government in the domain of urban policy (Jaillet, 2003), as all the previous Socialist governments had done since 1981. After all, urban policy was initiated by the Socialists. However, Jospin remained silent on urban policy, immediately proposed security-related measures towards the *banlieues*, and adopted a discourse that shared many elements of the republican consensus and the right's discourse on the *banlieues* ('right to security', 'outlaw areas', 'republican values'). Before even the assignment of a Minister for the City (which had to wait 1998), new security measures directed mainly towards the urban policy neighbourhoods were already in place.

More security measures directed towards the so-called 'sensitive neighbourhoods' followed in 1999. The Interior Security Council, creation of which was announced by Jospin at the Villepinte colloquium in 1997, was formed the same year.[17] The new Minister for the City, Claude Bartolone, was present at two meetings of the council in 1999, although he was not formally a member of it. The first of these meetings took place in January

1999. In his speech at the press conference, Jospin stated that insecurity affected 'the values of the republic', and pointed to the need to conceive a global policy to 'act against violence' and its causes. 'These causes are known', Jospin continued: 'economic distress based on unemployment and precariousness; the social run-down of some neighbourhoods; the lack of integration of a part of youth of immigrant background.' Jospin defined three priorities. One of these was fighting against violence in school. The other two concerned 'the neighbourhoods'. The first was 'ensuring an effective presence in neighbourhoods and sensitive places'. The measures to this aim involved the recruitment of 7,000 new police and gendarme, the creation of 30 new MJDs and a '*police de proximité*' (literally, 'proximity police') in 'sensitive areas'.[18] The second was 'improving the efficiency of responses to acts of delinquency'. Young people and minors were of particular concern: 100 'reinforced education centres' – 77 more than was originally planned – would be created for 'minors for whom being taken out of their usual environment for a few months is necessary'. For the most 'difficult' ones, 50 detention centres would be created in order to put them 'at a distance from their neighbourhoods' while awaiting judgement.[19] The measures proved effective: the incarceration of minors doubled in less than three years (*Marchands de sécurité*, 2002).

The council met again in April 1999 to discuss, among other issues, the relationship between urban policy and the CLSs (Local Security Contracts), launched in October 1997. This was part of the preparations for the so-called 'new generation of City Contracts' that the government was planning to start the following year. The council decided that the CLS would be 'one of the essential components of future city contracts'.[20] As Minister for the City stated in another context, the urban policy neighbourhoods were, and would be, given priority in terms of policies of security, including the *police de proximité*, the CLSs and the MJDs. Two hundred and nineteen districts would have the *police de proximité* starting from January 2002. By June 2001, 527 CLSs, of which 60% were in the intervention areas of urban policy, were signed, and another 209 were under elaboration (Cour des Comptes, 2002).[21] In 2002, there were 89 working MJDs in metropolitan France, and 38 were expected to be constructed by the end of the year.[22] Figure 5.1 shows the location of MJDs in metropolitan France in 2002. Not all of them correspond to communes involved in urban policy. Some are in small communes whose remoteness from the court on which they depend justifies the presence of MJDs (for example, Lannion in Brittany, Brive-la-Gaillarde in southern Limousin, Bergerac in Aquitaine). However, 77 out of 89 MJDs are located in communes with urban policy neighbourhoods.

Security, however, was not the only preoccupation of the Jospin government, although it replaced urban policy as a priority issue compared to the

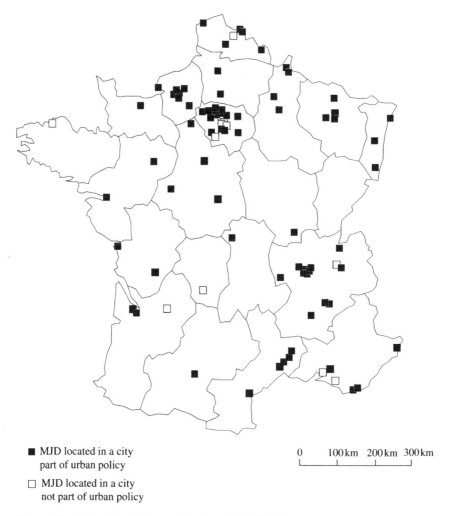

■ MJD located in a city
   part of urban policy

☐ MJD located in a city
   not part of urban policy

0   100 km   200 km   300 km

Figure 5.1   The location of *Maisons de Justice et du Droit* (MJDs) in mainland France in 2002 (*Source*: Ministry of Justice)

previous Socialist governments. A year after coming to power, the government started working on the orientations of urban policy.

Remaking Urban Policy in Republican Terms

On 30 June 1998, the inter-ministerial committee for cities (CIV) met under the presidency of Prime Minister Jospin. The press release after the meeting stated that the government had decided upon its orientations regarding urban policy, based partly on the propositions of the Sueur Report of the same year. Referred to as 'a new ambition for cities', these orientations consisted of four objectives: 'to guarantee the republican pact on the whole of territory; to ensure the social cohesion of our cities; to mobilize around a collective project; and to construct a new democratic space with inhabitants' (CIV, 1998a).

The first objective was conceived to address 'the urban crisis', which was seen first of all as a consequence of social and economic transformations. This 'urban crisis' was characterized by high levels of unemployment and 'violence and insecurity', which not only challenged the 'right to security', but also hampered 'the integration of the neighbourhood into the city'. The role of urban policy, in this sense, was defined as 're-founding the republican pact'.

'Ensuring social cohesion' was the second objective defined by the CIV. This implied urban renewal programmes aimed at the physical transformation of several neighbourhoods, and housing policies aimed at achieving 'social and urban mixity'. However, neither 'social cohesion' nor 'mixity' (social or urban) was explained. The CIV presumed that more 'mixity' would lead to 'social cohesion', establishing, therefore, a causal link between an undefined cause and an undefined effect.

The third objective was 'mobilizing around a collective project'. This implied cooperation between different territorial collectivities (regions, departments and communes). Inter-communal cooperation was particularly emphasized. The state's role would be to ensure the equality of public services in order to 'guarantee the respect of republican values and principles'.

The last objective, 'constructing a new democratic space with inhabitants', followed a long tradition in urban policy: that is, reiterating the importance of the participation of inhabitants and criticizing previous programmes for the lack of it. The CIV declared that the state would not sign contracts unless forms of participation of inhabitants were specified. Defining these conditions, however, was not binding. Although the 'involvement' of inhabitants was encouraged, even required, they were not officially recognized partners in the process of contract preparation.

The CIV met for the second time in less than six months, again under the presidency of Jospin, to further elaborate the 'new ambition for cities', and to clarify the features of the 'new generation of city contracts' for the period 2000–6. Once again, the state's determination to 'guarantee social cohesion and the respect of republican values' was affirmed, and urban policy's three priorities were defined as employment, security and education (CIV, 1998b). A few weeks after this meeting, Jospin issued a circular addressed to prefects in order to clarify the features of the 'new generation of city contracts' for the XIIth Plan (2000–6).[23] He emphasized that while interventions targeted at priority areas would continue, the contracts were to be conceived at the scale of commune and agglomeration, with the general objective of 'fighting urban and social segregation'.

Thus started the Jospin government's actions in the domain of urban policy, which included an impressive amount of measures. What the Jospin government did in this period was more or less a remake of the measures of late 1980s and early 1990s – extended and couched in 'republican' terms. It reintroduced the City Contracts programme (which, as we have seen in Chapter 3, was initiated in 1989), passed two laws that sought to encourage inter-communal cooperation and change the scale of intervention to the agglomeration (these ideas were already there in the late 1980s with the City Contracts programme and the shift from DSQ to DSU), another one that was aimed at a more balanced distribution of social housing (just like the LOV of 1991 which we have seen in the previous chapter), and introduced a large-scale physical renovation programme, which took after the *Grands Projets Urbains* (GPUs) of the early 1990s. Similarities did not end here: all of these were preceded by another law that was passed in 1998 to fight against exclusion, as was the case in 1988 when the law introducing a minimum insertion income (RMI) was passed with the same stated objective. The 'left hand' of the state was catching up with its 'right hand'.

In spatial terms, these measures implied an opening up of the spatial focus of urban policy, which was severely limited by the statist geography of the *Pacte de Relance*. The reintroduction of the contracts led to a process open to negotiations between the state and local collectivities rather than a geography imposed by the technocrats of the central state. This, however, did not mean an abrupt end to the geography and measures of the *Pacte de Relance*. The system of tax concessions in designated areas (ZFUs) would continue, but would be phased out and would not be renewed after 2002. In the meantime, measures were introduced for 'moralizing' this procedure (such as the suppression of tax concessions for firms moving from one area to the other; see DIV, 2000b).

One of the objectives of the previous generation of City Contracts was to encourage inter-communal cooperation, and to change the scale of urban policy from the neighbourhood to the larger city-region, while addressing the problems of particular neighbourhoods. The Sueur Report of 1998 had emphasized this issue as well, arguing that the proper scale for urban policy should be the agglomeration, which is constituted by the central city and its *banlieues*. The new generation of City Contracts took this issue into consideration. It was complemented by two laws, both of which tried to encourage the opening up of the scale of urban policy, and inter-communal cooperation. The 'Voynet Law' of 25 June 1999 gave the agglomeration an official status, and sought to encourage contracts conceived at the level of agglomeration.[24] City Contracts could be integrated into these contracts as their 'territorial and social cohesion' facet. The other law, the 'Chevènement Law' of 12 July 1999 was aimed at encouraging inter-communal cooperation through a redistribution of business tax between communes provided they organized in the form of an 'Agglomeration Community' ('*Communauté d'Agglomération*').[25] These communities would bring together communes to include at least 50,000 inhabitants (which was a requirement to be eligible for a contract in the framework of the 'Voynet Law'), and would have a juridical status. With this law, communes organized in the form of agglomeration communities would receive more money from the state for their functioning, but also see their responsibilities increased in the domains of economic development, planning, transportation, housing and urban policy.

These two laws were complemented by another aimed at achieving a more balanced distribution of social housing. Passed in 2000, the Solidarity and Urban Renewal Law (*Loi de Solidarité et Renouvellement Urbains*, SRU) had the same objective as the LOV of 1991: a more balanced distribution of social housing by requiring communes to attain an objective of 20%.[26] However, its scope was larger than the LOV's: whereas the LOV obliged communes in agglomerations with more than 200,000 inhabitants to provide 20% social housing, the SRU required the same proportion of social housing from communes of more than 3,500 inhabitants (1,500 in the Paris region) in agglomerations with more than 50,000 inhabitants in 20 years. Failing to do so would incur a fine of 1,000 francs (152 euros) for each 'missing' social housing unit per year.

The communes, however, did not want social housing because they did not want immigrants. They preferred simply to pay the fines they incurred for not achieving the required amount of social housing. The mayor of Vaulx-en-Velin, Maurice Charrier, commented on the SRU as follows:

Well, all right, one day I put myself in the place of a mayor of the west of Lyon, and I took the case of a commune, it was St-Foy-lès-Lyon. See, I had worked out that, in fact, not respecting the SRU law obliged me to pay one million francs . . .

Fines?

. . . in fines, by year. After ten years, I think! I'm convinced that my population is ready to have taxes raised by one or two percent as long as we have no social housing [which means immigrants, as it becomes clear later in the interview].[27] See? [. . .] On the other hand, if you have to build 500 social housing units, you'll have to build a school. And just that school will cost more than a million in maintenance. (Interview, Maurice Charrier)

In this sense, the SRU did not push hard enough, which was also a critique directed at the LOV. Following the autumn 2005 revolts, the SRU and its requirements in terms of social housing became topical issues again. The SRU's requirement of 20% social housing was to be achieved in 20 years, and an assessment would be made every three years starting from 2002 (when the law came into effect). At the end of 2005, therefore, an assessment was made by the prefects to see if communes concerned by the SRU had satisfied their three-yearly obligations, and to fine those that failed to do so. The situation was as follows: among 735 communes concerned by the SRU, only half of them satisfied their three-yearly requirements. Among the other half, 200 of them were so behind the required amount that their fines would be augmented. Some of the smaller communes (fewer than 10,000 inhabitants) did not construct social housing at all, mainly for electoral concerns. In the Paris region, fewer than half (91) of the 186 communes concerned satisfied the requirements. Among those who were behind the required amount, one quarter did not construct social housing at all (*Libération*, 7–8 January 2006: 14). Shortly before these results were announced, Minister of the Interior Sarkozy was ironical about the SRU: 'What failed in Eastern Europe should work in France, so the aim of a social housing policy should be to cover France with HLM!?!' (*Le Parisien*, 4 December 2005; cited in *Le Canard enchaîné*, 1 February 2006: 5). Sarkozy was the mayor (between 1983 and 2002) and is, at the time of writing, the deputy mayor of Neuilly-sur-Seine, located to the west of Paris, between the city limits and La Défense. The proportion of social housing in this commune is 2.6% (*Le Canard enchaîné*, 1 February 2006), and in 2005, only half of the required three-yearly amount was constructed. Thus the commune, one of the wealthiest *banlieues* of the Paris region, would be fined 57,152 euros plus an additional 14,860 euros for falling that behind the required amount (*Libération*, 7–8 January 2006: 14).

Nevertheless, the SRU, together with the 'Voynet Law' and 'Chevènement Law', was an important step for opening up the spatial focus of urban policy by encouraging contracts conceived at the scale of agglomeration. All three laws had a redistributional aspect; they were all marked by an emphasis on solidarity (Goze, 2002), although they also implied more responsibility and bidding for communes, interpreted as a sign of the relative disengagement of the state (Donzelot, 2006; Le Galès, 2005). Indeed, the Jospin government's discourse on urban policy was characterized by elements of republican rhetoric, solidarity being one, social cohesion the other. This was a departure from the discourse of the early years of urban policy, where republican rhetoric was largely avoided in an attempt to connect issues to structural dynamics with the use of terms such as 'inequality' and 'social justice' (see Silver, 1994). The Jospin government's emphasis on 'social cohesion', however, implied an organic conception of society with no structural conflicts. Its emphasis on 'solidarity', on the other hand, implied a tacit acceptance of lasting inequalities (see Jobert and Théret, 1994; Schmidt, 2002).

Urban policy was on its way to 're-found the republican pact', guarantee 'the right to security', and ensure 'social cohesion' when an eminent government institution questioned its *raison d'être*.

Whither Urban Policy?

Jospin's 1998 circular regarding the new generation of city contracts had identified the general objective of urban policy as 'fighting urban and social segregation'. With this in mind, more specific programmes were defined as follows: mixity in housing; diversification of functions in social housing neighbourhoods; coordination of the interventions of different actors involved in actions aimed at prevention and fighting exclusion; employment and local economic development; opening up of neighbourhoods through a coherent transportation policy; equality of citizens before public services (education, health, culture, justice, access to rights); prevention of delinquency; public tranquillity; and the integration of immigrants and their families.[28]

The list was ambitious, if not excessive, and begged the question as to whether urban policy could reasonably be expected to deal with all these issues. Despite the engagement of the state for years, the financial means of urban policy were still token compared to the money spent on its neighbourhoods in the framework of social service and welfare provision (Béhar, 1998; Le Galès, 1995). And yet, urban policy was the most publicized of them all, enjoying a status of priority for years in successive governments.

The list also evoked another question about the nature of urban policy. What, exactly, was this policy about? This question lay at the heart of a report that appeared in February 2002, prepared by a 'heavy-weight' government institution, the *Cour des Comptes* (equivalent of the Government Accounting Office in the United States or the National Audit Office in the United Kingdom).

The *Cour des Comptes*'s (the *Cour* hereafter) critique started with a questioning of the denomination that was used to refer to the interventions of the state in the neighbourhoods – '*la politique de la Ville*', a policy for *the City*. This denomination was rather ambiguous, it was argued, since the interventions did not apply to all cities, or to the whole city. It was used to refer to actions that were concerned with many issues (remember Jospin's list) – from physical planning to social issues, from cultural activities to political preoccupations, from economic development to 'public tranquillity', from housing to the 'integration of immigrants – but nothing specific. For the *Cour*, this ambiguity signalled deeper problems: 'The ambiguity of the expression and the difficulty of defining it [i.e. urban policy] simply reveal deep-seated problems about the content and the objectives of a policy which is, however, considered a priority by successive governments' (Cour des Comptes, 2002: 7).

The *Cour*'s report criticized the imprecision of the objectives of urban policy, which it saw as the consequence of the constant broadening of its intervention areas and the issues it sought to address. Urban policy 'until now has been characterized by the imprecision of its objectives and of its strategy, and by a quest for public exposure [*une volonté d'affichage*] which means that new measures are constantly being devised' (2002: 213). This critique brings to mind the comments of Nordine, one of the founding members of the MIB (*Mouvement de l'Immigration et des Banlieues*) whom we 'met' in the previous chapter:

> [Urban policy is] a providential windfall for local authorities and others, though not necessarily used appropriately for inhabitants. This does not mean that this money was diverted, but simply that something totally different was done with it. As far as we're concerned, unless someone can show otherwise, words were all we got [*ça a toujours été des effets d'annonce*], up until now and I believe that urban policy is pure make-believe [*poudre aux yeux*] for the inhabitants. (Interview, Nordine, MIB)

The *Cour*'s critique was informed by case studies of seven sites, which were referred to in the report as the 'historical' sites of urban policy. All but two of these sites have been included in urban policy since its inception, benefiting from various measures aimed at its so-called 'priority neighbourhoods'. These sites included the social housing neighbourhoods of the following communes, all in the *banlieues*:

In the Ile-de-France region:
* Mantes-la-Jolie (included since 1982; revolts in 1991);
* Grigny (included since 1982; revolts in 1995 and 1999); and
* Clichy-Montfermeil (included since 1989; starting place of the autumn 2005 revolts).

In the Rhône-Alpes region:
* Vénissieux (revolts in 1981; included in 1982; revolts again in 1991); and
* Vaulx-en-Velin (included since 1984; revolts in 1990 and 1992).

In the Provence-Alpes-Côte d'Azur region:
* Marseille (revolts in 1981; included in 1982 and 1983, northern neighbourhoods first, and the centre the following year).

In the Nord-Pas-de-Calais region:
*   Roubaix/Tourcoing (revolts in Roubaix in the early 1980s; included in 1982; Tourcoing included in 1989, revolts in 1992 and 1993).

Let us take a look at the situation in these neighbourhoods in order to comprehend the basis of the *Cour*'s severe critique of urban policy.[29] All of these sites were beneficiaries of various urban policy programmes. All of them were designated as ZFU following the *Pacte de Relance* of 1996, except Vénissieux, which became a ZFU in 2003. The ZFUs were conceived with the aim of creating jobs for the inhabitants living in these areas, which were made attractive for business through tax concessions. The unemployment rates in these areas, however, continued to increase. From 1990 to 1999, the unemployment rate in the ZFU of Mantes-la-Jolie rose from 15.7% to 25.7%; of Grigny from 17.0% to 26.2%; of Clichy-Montfermeil from 17.6% to 27.9%; of Roubaix/Tourcoing from 26.9% to 34.4%; of Vaulx-en-Velin from 17.9% to 28.4%; and the ZRU of Vénissieux (which is now a ZFU as well) from 20.3% to 29.1%.[30]

These unemployment rates followed regional patterns in a much more amplified way. For the same period, the unemployment rate in Ile-de-France increased from 8.5% to 11.5%; in Nord-Pas-de-Calais from 15.1% to 17.6%; and in Rhône-Alpes from 8.9% to 11.0%. The most hard-hit groups were young people aged between 15 and 24 and foreigners, with unemployment rates reaching as high as 46.7% for the former and 43.7% for the latter (Roubaix/Tourcoing), and never below 37% for the former and 28% for the latter.

The figures were similar when the entire list of neighbourhoods was taken into consideration (Table 5.1). From 1990 to 1999, unemployment rate in the priority neighbourhoods of urban policy increased by 35%. In 1999, one quarter of the active population in the priority neighbourhoods of urban policy was unemployed.

As Table 5.1 shows, the proportion of foreigners is twice as high in the priority neighbourhoods of urban policy as in their agglomerations, and thrice the national mean. This has to do with the allocation of social housing and the history of these neighbourhoods, most of which were constructed in the *banlieues* with concentrations of social housing estates (in 1999, 61% of the inhabitants of priority neighbourhoods lived in HLMs). As we have seen in Chapter 3, when the middle-class populations started to leave the HLMs towards the end of the 1970s, many immigrants had moved into these areas, partly because they were not welcome by the property market in the central city, or pushed away from the city centres because of urban renewal programmes and speculation.

The proportion of young people is also higher than regional and national means, and they seem to be the most hard-hit group by increasing

Table 5.1 Characteristics of priority neighbourhoods of urban policy in comparison with cities and metropolitan France, 1990–9 (%)

	Priority neighbourhoods of urban policy		Agglomerations with priority neighbourhoods		France total	
	1990	1999	1990	1999	1990	1999
Unemployment rate	18.9	25.4	11.5	14.3	10.8	12.8
Unemployment rate/15–24	28.5	39.5	20.7	27.0	19.9	25.6
Unemployment rate/foreigners	26.2	35.3	19.8	25.8	18.8	24.1
Proportion of young people/15–24	18.2	16.2	15.8	14.2	15.0	13.0
Proportion of foreigners	18.6	16.5	9.0	7.9	6.3	5.6
Proportion of people with no high-school diploma (excluding students)	39.3	33.1	26.8	18.7	29.1	20.0

Source: INSEE-DIV (no date).

unemployment. In 1999, four out of ten young people living in the priority neighbourhoods of urban policy were unemployed. This may have to do with lower levels of education in a context in which there is less demand for unskilled labour, although the table suggests an improvement in educational achievement. There may also be other factors, such as discrimination in the job market because of ethnic origins or of being associated with a stigmatized neighbourhood. The interviews from Vaulx-en-Velin, as we will see in the following chapter, suggest such a possibility.

Although unemployment rates increased throughout France following the crisis of the 1970s, the priority neighbourhoods of urban policy seem to have been hit particularly hard by this increase. The reasons for such increases in unemployment are no secret. The economic instability that ensued from the oil crisis in the 1970s was influential in increasing unemployment rates. The major change, however, was brought about by the economic restructuring processes of the 1980s and 1990s, which translated into sharp declines in the manufacturing sector after the relocation of firms in parts of the world that were economically more profitable. There were

4.6 million people employed in the manufacturing sector (construction not included) in France in 1989, and half a million of these jobs were lost from 1989 to 1994. This trend was aggravated even further with technological advances and the development of new service sectors, increasing the demand for more skilled labour than was already available following the losses in the manufacturing sector. Many working-class neighbourhoods, most of which today are urban policy's priority neighbourhoods, were hit severely by unemployment ensuing largely from plant closures in manufac- turing and industrial sectors, and the restructuring of demand for relatively more skilled labour (OECD, 1998).

French urban policy, as the *Cour*'s report held, was not 'efficient' enough to address these intensifying problems. Or, at least, it was not capable of delivering what it kept promising increasingly since the 1990s, which was the creation of jobs for the inhabitants of priority neighbourhoods (although firms were created following tax concessions). Unemployment remains a major problem in the urban policy neighbourhoods, although the discursive articulations of these neighbourhoods, as we have seen, have changed remarkably, from social development of neighbourhoods in the 1980s to the 'republican pact' in the 1990s. The coming to power of an even more authoritarian and 'republican' government would bring about further changes and increase the pressure on the police order.

The Police Order and the Police State

The Jospin government's last year and the early years of the right-wing government that replaced it were marked by the repercussions of Septem- ber 11. Two months after the attacks, the Jospin government passed a law on 'security in daily life', known as the 'LSQ Law'.[31] This law sought to associate mayors with security actions, extend the possibilities for police checks, and made it possible for the police to intervene in the common areas of buildings, which basically was aimed at youth gathering at building entrances in social housing estates. Jospin had already announced that the issue of security was a priority for his government. The more the presiden- tial elections of 2002 approached, the more 'securitarian' Jospin (who was a candidate himself) became. Indeed, with the exception of the extreme left and the greens, every single candidate prioritized the issue of security in their campaign. So much so that a couple of months before the election, Chirac was urged by his fellow party members to 'toughen up' his discourse on security, which, to start with, was not weak at all with terms such as 'zero tolerance' and 'zero impunity' (see *Le Monde*, 27 February 2002a: 10). Jospin lost eventually, and Chirac – a recidivist himself enjoying immu- nity as the President of the Republic – was re-elected President after the second round of elections, where his opponent was the extreme right leader Le Pen.

The Raffarin government came to power in 2002 in this context to fulfil such promises, with Nicolas Sarkozy as the Minister of the Interior. Sarkozy immediately passed two laws, one only a few months after taking office ('Sarkozy Law'), and another complementary one a year later ('LSI' on interior security).[32] With these laws, Sarkozy gained control over the police and gendarmerie with a stated aim to repress 'new forms of delinquency'. These laws made identity checks and vehicle searches almost arbitrary, justified merely by 'reasonable ground for suspicion'. They targeted four categories of people in particular, introducing repressive measures towards them: prostitutes, beggars, gypsies and youth gathering at building entrances (Jean, 2004). The Jospin government had also targeted youth gathering at building entrances with the LSQ a year earlier, making it possible for the police to intervene. The LSI pushed this measure to the extreme, and turned such gatherings into a criminal offence (provided they were deemed to hinder free movement of others) punishable by a penalty of two months in prison and a fine of 3,000 euros. This is how Guy from MIB commented on these developments:

> This is colonial, we've come back to dynamics of spatial control, which belong to the domain of colonial policies. We could say it is neo-colonial; obviously things are not taking place in the same way, but there are still traces of that, not everywhere but definitely in the [social housing] neighbourhoods. (Interview, Guy, MIB)

Complementary to Sarkozy's laws were Minister of Justice Dominique Perben's two laws, the first one passed in September 2002 ('Perben Law I') and the second in March 2004 ('Perben Law II').[33] While Perben Law I intensified penal responses to minor delinquency, Perben Law II increased police custody to four days, and extended the scope of the notion of 'organized gang', making it possible to treat infractions of this kind by using measures hitherto limited to terrorism (Jean, 2003; Nuttens, 2004). In this sense, it is significant that Sarkozy insisted on the 'organized' nature of the revolts of autumn 2005, a claim refuted by a report prepared by the French Intelligence Service (which eventually cost its director his job).

The repressive measures introduced by these laws are only partially presented here, with a focus on measures that relate more closely to *banlieues* and youth (for more on these laws, see Bourmeau, 2003; Brun, 2002; Jean, 2003, 2004; Nuttens, 2004). These laws, as Jean (2003: 28) argued, were unprecedented in terms of reinforcing the powers of the police and the penalization of society.[34] One of the principal targets of these new laws was the *banlieues* and *banlieue* youth (Bonelli, 2003; Wacquant, 2003). As Sarkozy had announced immediately after taking office in May 2002, one of the main targets of his law and order policies would be what he referred to as 'sensitive neighbourhoods' and 'outlaw areas' ('*zones de non-droit*'). One of his immediate measures was the distribution of flash-ball guns to the proximity police working in 'sensitive neighbourhoods', although the measure was criticized by many human rights associations as a provocation of the *banlieue* youth and a departure from the main mission of the proximity police, which originally was prevention. Flash-ball guns fired rubber balls, and were used since 1995 by the anti-criminal brigade only (*Le Monde*, 18 May 2002; 31 May 2002b). Thus started Sarkozy's offensive with his stated conviction that 'repression is the best of preventions' (*Libération*, 11 July 2002: 2). For Michel Tubiana, the president of the League of Human Rights, these developments were '[t]he worst step back for human rights since Algeria' (*Libération*, 25 May 2004).[35] In the 2004 annual report of the League, the actions of the government were interpreted as follows: '2003 was a dark year for liberties. Seldom in the history of the Republic did any government as rapidly after its accession of power set up, to accompany its regressive social policies, a system as efficient to restrict the citizens' guarantees' (Leclerc, 2004: 27).

Back to the Statist Geography

'After measures aimed at restoring the republican order, the fight against injustice has started,' declared Jean-Louis Borloo, the new Minister for the City, soon after the passing of first the Sarkozy and then the Perben laws (*Libération*, 3 October 2002). Shortly after taking office, Borloo had outlined his vision of urban policy, which he referred to as a 'new battle of France'. The 'republic', he had maintained, had to guarantee the protection of the inhabitants of the urban policy neighbourhoods through its 'forces of order' (*Le Monde*, 28 May 2002: 12). Borloo had identified three priorities for urban policy: 'break up the ghettos'; 'jobs and professional insertion'; and 'simplification of procedures' (DIV, 2004: 24). Thus, for the first time in urban policy, a Minister for the City explicitly and without reservations referred to the urban policy neighbourhoods as ghettos 'at the margins of national territory' (*Le Monde*, 19 June 2003).[36]

The priorities defined by Borloo guided a new law on urban policy, known as 'Borloo Law', passed in August 2003.[37] This law translated the first priority, 'break up the ghettos', into demolition programmes; the second, 'jobs and professional insertion', into more areas with tax concessions; and the third, 'simplification of procedures', into the creation of the 'National Agency for Urban Renovation' ('*Agence Nationale pour la*

Rénovation Urbaine', known as ANRU). Except for this last measure, therefore, the Borloo Law did not introduce any novel measures. What it did was radicalize measures that had already been introduced by the *Pacte de Relance* in 1996 and the Jospin government's urban renewal programme in 1999.

The Borloo Law was conceived along three main lines that reflected the three priorities identified above: physical renovation, economic development and institutional restructuring. The first, physical renovation, had the objective of achieving 'social mixity' (Article 6). The measure introduced to this end was the demolition of social housing estates. Thus, the law radicalized a measure that was already introduced by the Jospin government's urban renewal programme, which included demolitions, renewal programmes (*Opérations de Renouvellement Urbain*, ORUs) and the so-called *'Grands Projets de Ville'* (GPVs). In the framework of these programmes, conceived in relation to City Contracts, 50 sites were designated for large-scale urban projects (GPVs), and 30 for smaller-scale renewal projects (ORUs). The rhythm of demolition was announced first as 10,000 to 12,000 units per year (CIV, 1999), and as the 2002 elections approached, this number went up to 30,000 (CIV, 2001a). The Borloo Law radicalized these measures with a stated objective of 200,000 demolitions, 200,000 rehabilitations and 200,000 new social housing units in five years. This measure, however, worries inhabitants and associations, who fear that reconstructions will not equal demolitions in a context where there is already an acute social housing shortage. There are also signs that suggest that demolition decisions are imposed on inhabitants. In February 2005, among the 83 projects approved by the ANRU there were 39,000 planned demolitions and 36,000 planned reconstructions (*Libération*, 24 February 2005a: 6–7).

The second main line, economic development, had the objective of 'reducing territorial disparities' (Annex 1, section 2 of the law). The measure introduced to this end was the re-opening of the previous 44 ZFUs (which were not renewed by the Jospin government) and the creation of 41 more areas benefiting from tax concessions. Thus, the Borloo Law geographically expanded a measure introduced by the *Pacte de Relance* in 1996. Following the autumn 2005 revolts, 15 more ZFUs were created (CIV, 2006), raising the total number of ZFUs to 100 (seven of them in France's overseas territories). Opinions about the ZFUs, however, vary. A 2001 report pointed that they were not that helpful for providing jobs for the inhabitants of the designated areas, which was also the critique of the Sueur Report in 1998. Another report prepared in 2003 gave a more positive evaluation, stating that first 44 ZFUs had created 10,000 firms and 45,000 jobs (DIV, 2004). But there lay the problem with the ZFUs. New jobs were usually precarious, most of the time never turning into a stable position. Furthermore, there was no indication that jobs went to the inhabitants of designated areas. Although the firms were obliged to hire one third of the employees from the designated areas (as opposed to one fifth, as the *Pacte de Relance* had it), the evaluation was based purely on declarative data provided by the firms themselves, which was not all that reliable (see, for example, *Le Monde*, 18 February 2003 and *Libération*, 8 May 2006). However, the government created more ZFU, and is currently trying to stretch the boundaries of existing ones, open up the measure to large firms as well (currently it is limited to small firms with fewer than 50 employees), and make them more attractive for business in financial terms (*Le Monde*, 2 December 2005: 14).

The third main line, institutional restructuring, had the objective of simplifying procedures. To this end, the National Agency for Urban Renovation (ANRU) was created. ANRU would centralize, manage and distribute subventions, thus providing a single stop (*'guichet unique'*) for the financing of renovation programmes (Depincé, 2003: 28). This indeed was a significant reorganization, but the creation of ANRU meant much more than just simplification of procedures. The creation of ANRU, and the Borloo Law in general, re-centralized urban policy. This re-centralization implied an urban policy that risked becoming purely procedural, reduced to the physical transformation of the built environment or the designation of more ZFUs. It also took urban policy back to the statist geography of *Pacte de Relance* and remarkably limited its spatial focus.

The creation of ANRU for physical renovation, on the one hand, and re-opening and expansion of ZFUs, on the other, implied a major divergence from previous urban policies, which combined physical interventions and measures with social aspects. Urban policy had always had a social dimension. The Borloo Law brought about a dissociation in terms both of the nature of urban policy and of the relationship it had established between the state and local collectivities (Epstein, 2004; Epstein and Kirszbaum, 2006; Estèbe, 2004; Jaillet, 2003). Regarding the former, it narrowed down the engagement of the state to physical interventions only, leaving the social dimension to the initiative and capabilities of local collectivities. Regarding the latter, it dissociated the contractual relationship between the state and local collectivities, which also was a feature of the *Pacte de Relance*. The previous City Contracts programme was more flexible in terms of objectives and actions, which were negotiated by the state and local collectivities. The new programme, on the contrary, was very rigid; it defined objectives and actions by law (Annex 1 of the law precisely defined the objectives and indicators of urban policy actions over ten pages). Thus, the objectives of urban policy were given – not negotiated (Epstein and Kirszbaum, 2006). Furthermore, instead of a process of negotiation, a hierarchical relationship was established between a central agency (ANRU) and local collectivities, which risked reducing the latter's role to a purely administrative one executing the centrally defined actions (Estèbe, 2004).

In spatial terms, the Borloo Law implied a narrowing down both in terms of the geography urban policy consolidated through its spaces of intervention, and of the way in which spaces of intervention themselves were conceived. In terms of its geography, the Borloo Law, as noted, took urban policy back to the statist geography of *Pacte de Relance*. This geography was centrally defined – not negotiated – and absolute, constituted by neatly delimited spaces of intervention as if they existed in and of themselves. Thus, urban policy was re-focused on the 751 ZUSs that were defined in 1996, leaving out some 600 neighbourhoods that were defined as priority in City Contracts but were not part of the centrally designated ZUSs (Epstein, 2004).[38] The attempts to open up the focus of urban policy by encouraging projects conceived at the level of agglomeration thus came to an end. When asked his opinion of the SRU, which sought to achieve a more balanced distribution of social housing among communes, Borloo stated that it was a 'good idea', but that sanctioning communes that did not construct social housing was 'idiotic' (*Le Monde*, 28 May 2002: 12).

In terms of the conceptualization of spaces of intervention, the Borloo Law reduced the spaces of urban policy to an ensemble of housing estates. Over the years, urban policy gradually shifted focus from the inhabitants themselves to spaces of intervention. The Borloo Law followed this trend to the extreme, and transformed urban policy to a policy 'addressed not to inhabitants but to space seen as a problem in and of itself' (Estèbe, 2004: 255). Unsurprisingly perhaps, with the Borloo Law, the 'eternal *leitmotiv*' of urban policy – participation of inhabitants – disappeared. Not only was there no allusion to it in the law, but also the institutional structure introduced by the law closed down the spaces for the engagement of inhabitants – which were not all that open to start with. ANRU – not the inhabitants – was now the local officials' interlocutor, where they had to have their projects validated and give an account of their actions (Epstein and Kirszbaum, 2006).

The inclusion of inhabitants in the process – which takes time – would undermine the urban renovation programme introduced by the Borloo Law – which was indexed to urgency and efficiency rather than to democracy. Demolition was rapid and politically spectacular, even though it might not be the best of choices. In 2004, the Ministry of Culture commissioned three architects to devise ways other than demolition–reconstruction. They remarked that demolition was absurd for two reasons. First, it was absurd simply because there was a shortage of social housing (the waiting list for social housing consists of one million applications). Second, it was possible to transform buildings instead of demolishing them with a cost three to six times less. This, however, implied a process that involved the inhabitants themselves. It proceeded slowly as it followed the specificities of each case, and was not as spectacular as demolitions in terms of demonstrating the government's effort (*Libération*, 24 February 2005b: 7). Borloo wanted a programme that was visible and rapid. But it had to be efficient too. Thus, a national observatory was created as well in order to measure the efficiency of policies pursued in the ZUSs in relation to the objectives and the indicators defined in the law.[39]

Conclusions: Repressive Police

Urban policy after 2002 shared the same features as the same period's security policy: extreme measures focused on results with constant evaluation, and limited in scope, proposing immediately visible actions rather than comprehensive measures that address the structural causes of problems. With the Borloo Law, urban policy entered a new era with a more pronounced neoliberal orientation. Such an orientation was already discernible in the *Pacte de Relance* of 1996, but the social dimension was still there. The Borloo Law brought about a retreat of the state from the social in the domain of urban policy. Urban policy was reduced to a renovation policy, on the one hand, and to a policy of designating tax concession areas, on the other. Emphasis was put on urgency and efficiency, with a system of constant evaluation. In the process, the inhabitants were replaced with space, conceived either as the area between the lines delimiting tax concession areas or as an ensemble of housing estates that need to be demolished. This substitution was perhaps best exemplified by the remarks of Sarkozy cited by Borloo: 'There are *spaces* that have so many more disabilities that, if they are not given more, they will never make it' (Borloo, 2004: 105; emphasis added).

For Borloo, such territories had to be addressed by 'our republic' in order to reduce the 'territorial fracture' they induced, 'which has today become social, indeed ethnic' (Borloo, 2004: 105). Another feature of urban policy, not only in the post-2002 period but since the early 1990s, was this emphasis on the 'ethnic' nature of the problem of social housing estates in *banlieues*, seen to be threatening the integrity of the republic and its values. In this sense, 'the *banlieue*' became a 'useful' metaphor to talk about politically more fragile issues (Estèbe, 2004). It has become commonplace to bring up the same spatial references – '*banlieue*' or 'neighbourhoods' – when talking about issues such as the 'problem of integration of immigrants', the republican model of integration, social fracture, violence and insecurity. Furthermore, it has become possible to evoke all of these issues at once by a simple reference to 'the *banlieue*'. 'Ghetto' was the problem, 'social mixity' the solution. It was in the name of 'breaking up the ghettos' and achieving 'social mixity' that Borloo justified his demolition–reconstruction programme, which brought to mind the ill-famed destruction operations of the 1960s in the name of modernizing the old neighbourhoods of city centres (Jaillet, 2003). Massive demolitions of such neighbourhoods were eventually accompanied by reconstructions, but these operations, in the case of Paris, had diminished the population from 2.9 to 2 million, pushing away almost a million inhabitants to peripheral areas, which are today targeted by Borloo's demolition–reconstruction programme (Simon and Lévy, 2005). But the spectre of ghettos *à l'américaine* and the quest for 'social mixity' justified the new orientation of urban policy, which seemed to respond less – if at all – to the aspirations of the inhabitants of *banlieues* than to the 'fear' invoked by 'the *banlieue*'.

> Obviously, the quest for social mixity, which is used to justify the current policy of demolition–reconstruction, was not a response to claims from the inhabitants of those neighbourhoods, but a response to an anxiety from the rest of society taken up by politicians who wanted to believe that diluting problems would make them disappear, and called for a residential dispersal of visible minorities. (Epstein and Kirszbaum, 2006: 44)

But demolition was not the only response to the problem of *banlieues*. The penal state, as we have seen, has largely consolidated itself in and through the spaces of urban policy. There were already signs of the coming of the penal state in the early 1990s, with the engagement of the Ministry of Justice and French Intelligence Service with the question of *banlieues* and 'urban violence'. But the turning point was the 1997 Villepinte colloquium, which turned the issue of security into a consensual one. When urban policy was severely criticized by the *Cour des Comptes* report of 2002, the then Minister for the City, Claude Bartolone, had responded by stating that 'giving the *banlieue* a new face [was] the first objective to achieve'. This implied, he continued, 'first of all a safe environment':

> [S]ecurity has been our priority. Re-enforcement of police forces, development of proximity police, creation of *Maisons de Justice et du Droit* [MJDs], targeted actions of the police to fight all sorts of traffics: the decisions taken by the government to fight insecurity and violence benefit in the first instance the inhabitants of working-class neighbourhoods. Urban policy contributes to this fight against violence and criminality by funding prevention actions carried out within the framework of the local security contracts [CLSs], and the development of mediation. (Bartolone in Cour des Comptes, 2002: 266)

Bartolone's response summarizes, in a sense, the transformation of urban policy in the 1990s with a shift towards security. While the left talked more about prevention and the right more about repression, urban policy was marked by the issue of security in the 1990s. Whereas the urban policy neighbourhoods were sites of political experimentation in the early 1980s, starting with the mid-1990s they have become sites where the republican penal state has consolidated itself. Through the consolidation of this police order with discursive articulations that depicted an image of 'the republic' under threat by 'communities' and 'urban violence', the status of *banlieues* has shifted remarkably. The 're-foundation of the republican pact' implied, even required, more repressive measures towards *banlieues*, now standing for the spatially reified form of the 'threat' to the integrity and values of the republic.

Chapters in this part had the aim of charting the changing articulations of the spaces of urban policy. As we have seen, urban policy has consolidated and intervened in what I have called, non-pejoratively, a 'police order', which has included the same neighbourhoods for years. But the discursive articulations of this police order through the state's statements have changed remarkably over the years. The result was the consolidation of a police order almost in the literal sense. The republican penal state, the coming of which, as we have seen in the previous chapter, was already signalled in the early 1990s, consolidated itself mainly in and through the social housing neighbourhoods of *banlieues*, constituting the *banlieues* increasingly as 'badlands' that do not quite fit in the 'republican' imagery. As I have maintained in the first couple of chapters, however, this articulation as badlands through the state's statements is not an uncontested process. There are alternative voices formulated in *banlieues* which challenge the place assigned to them in the police order, and show that what the state's statements constitute as badlands are also sites of political mobilization with democratic ideals. So let us now move onto one of these 'badlands' and see what a group of local activists have to say.

Part III

Justice in *Banlieues*

6

A 'Thirst for Citizenship': Voices from a *Banlieue*

Vaulx-en-Velin between Official Processions and Police Forces

In his contribution to a report prepared by the National Council of Cities on the media and the urban policy neighbourhoods, Jean-Jacques Bozonnet, then responsible for the 'society' section of *Le Monde*, stated that journalists visit these neighbourhoods either 'when things go wrong, in the wake of police forces, or when things are better, in the wake of official processions for an inauguration' (CNV, 1991b: 45). The inhabitants of Vaulx-en-Velin had two such visits, in the autumn of 1990, with an interval only of one week. On Saturday, 29 September, the neighbourhood of Mas du Taureau in Vaulx-en-Velin was, in the words of the mayor, 'having a house-warming party'. Three years of rehabilitation projects were completed, and the so-called 'New Mas' was inaugurated with the participation of inhabitants, politicians and journalists. Furthermore, the inclusion of this *banlieue* in the following period of urban policy programmes for the Xth Plan (1989–93) was affirmed. Vaulx-en-Velin had been included in urban policy programmes since 1984, and it was seen as one of its 'exemplary' sites, as its mayor, Maurice Charrier, pointed out (1991–2: 92–3).

The next Saturday, 6 October, the New Mas was in flames, following the killing of a young inhabitant of the neighbourhood, 21-year-old Thomas Claudio. Claudio was killed after a police car cut ('deliberately', according to eyewitnesses) in front of the motorcycle his friend was riding. The rider lost control of his motorcycle, and Claudio, with no crash helmet, was fatally injured after having been thrown out from his seat. This is the version provided by two eyewitnesses and the rider himself, who survived the accident. As the police version goes, the rider lost control and the motorcycle glided and ended up running into a police car that happened to be there at that precise moment. There was not, however, much reasoning over which version was true. The relations between the police and the young inhabitants of the neighbourhood had always been very tense, and the death of the young man was 'one death too many', as many inhabitants of the neighbourhood put it.

That was the triggering incident – not of an uncommon sort, as we will see in the next chapter, on the *banlieue* revolts. Revolts of a hitherto unprecedented scale ensued, lasting for five days. Shops were sacked, and the newly built commercial centre was set on fire. There were wild confrontations between the riot police and the young inhabitants of the neighbourhood, who were joined by other young people from some of the neighbourhoods in proximity. 'There were 200–300 rioters [. . .], we had never observed incidents that serious in the *banlieues*' (interview, Lucienne Bui-Trong, French Intelligence Service). The shock following the incidents was unprecedented as well. The daily newspaper *Libération*, a week after the incidents, published a special section entitled 'Why Vaulx-en-Velin?' with the aim of informing its readers as to 'what happened in this *banlieue* presented as a model of rehabilitation'. The editorial piece opened with the following observation: 'Until the first Saturday in October, Vaulx-en-Velin seemed like a dream. For the Socialist parliamentary representative of the district, Jean-Jacques Queyranne, for the Communist mayor Maurice Charrier, and for the president of the urban agglomeration, Michel Noir from RPR, the rehabilitation undertaken in this *banlieue* of Lyon since 1985 was *simply exemplary*' (*Libération*, 13–14 October 1990: 23; emphasis added).

The Socialist government was once again taken by surprise. Prime Minister Michel Rocard charged the two vice-presidents of the National Council of Cities with a mission to visit Vaulx-en-Velin and prepare a report. A few weeks after the incidents, the report was ready. This rather short report had one major message to deliver: that the existing urban policy programmes were necessary, and they had to be pursued for the development of 'a new relationship of citizenship': 'If Vaulx-en-Velin is to become a reference, it should be about the modernization of the state, about a new relationship of citizenship' (CNV, 1990b: 2). And the neighbourhood of Mas du Taureau had important resources for the development of this new relationship of citizenship, thanks to the Social Development of Neighbourhoods (DSQ) programme: 'The DSQ policy pursued had a remarkable effect on the development of associative life in the neighbourhood of Mas du Taureau. This asset should allow for a rapid re-establishment of social dialogue' (CNV, 1990b: 3).

What, then, happened in this neighbourhood after the incidents? In this chapter, I tell the story of *Agora*, a politically engaged association created by the immigrant youth of the neighbourhood soon after the incidents with the intention to 'participate in the life of the neighbourhood'. *Agora* is of particular significance both socially and politically in Vaulx-en-Velin (Chabanet, 1999; Essassi, 1992), and remains one of the most prominent political associations in the *banlieues* of French cities. The story of *Agora* specifically illustrates many of the problems faced by *banlieue* inhabitants – youth in particular. It also provides insight into the local reverberations of urban policy as Vaulx-en-Velin has been included in urban policy since 1984, and *Agora* has been, since its creation in 1991, deeply involved in urban policy. Furthermore, the story of *Agora*'s political engagement illustrates the difficulties encountered by political mobilizations in the *banlieues*, exemplifying the limits placed upon them by 'the police'. As I have argued in Chapter 2, drawing on Rancière, one of the consequences of the consolidation of the police as an order of governance is that it creates a particular *locus of enunciation* for the governed. In this chapter, we will see the repercussions of the police through the story of *Agora*'s political engagement.

Before presenting the story of *Agora*'s political engagement, however, some background information on Vaulx-en-Velin will be helpful to get an insight into the economic dynamics that have affected this *banlieue* in the post-war period. This information should also give a sense of the trajectory of many of the social housing neighbourhoods in *banlieues* that are included in urban policy programmes, which followed a similar pattern of decline following the economic crisis of the 1970s and the economic restructuring processes that followed.

Vaulx-en-Velin after the *trente glorieuses*

Vaulx-en-Velin is part of the Lyon agglomeration in the department of Rhône, the region of Rhône-Alpes (see Figures 3.2, 3.3 and 3.4). Located in the east of Lyon, it has, like many of its neighbouring communes, a former Priority Urbanization Area (ZUP, hereafter). In this regard, the development of Vaulx-en-Velin is characteristic of many of the social housing neighbourhoods constructed in the peripheral areas of large cities during the post-war growth era: designation as a ZUP in order to respond to housing shortage, construction of social housing (HLM) mostly in the form of *grands ensembles* (large housing estates), sudden demographic expansion, major job losses following the crisis in the 1970s and economic restructuring in the 1980s, and gradual degradation of the housing stock due to poor quality of the buildings (cheap and rapid construction) and poor maintenance.

The demographic expansion of Vaulx-en-Velin had started before the construction of the ZUP, and was closely linked to immigration. For example, towards the mid-1960s, many *grands ensembles* were constructed in the neighbourhood of La Grappinière in order to accommodate repatriated settlers from Algeria. Later, families who were pushed away from the central areas of Lyon or neighbouring Villeurbanne because of speculative renovation projects settled in the commune, where they could find housing with reasonable conditions and prices.

The construction in the ZUP, of which Mas du Taureau neighbourhood, where the revolts had started, was a part, began in 1970, and 8,300 housing units were constructed in the following ten years. The ZUP changed the demographic structure of the commune dramatically. In a period of ten years, the population increased about 65% (from 26,837 in 1972 to 44,113 in 1982). After this quick expansion, the population of the city was more or less stabilized (44,153 in 1990 and 39,128 in 1999), with half of it living in the (former) ZUP. This demographic growth made Vaulx-en-Velin the fourth largest commune in the department, and the ninth in the region.

In the departmental rankings, Vaulx-en-Velin was the poorest commune with regard to per capita income in 1992 and 1993. Its demographic expansion was in large part made possible by the availability of industrial and manufacturing jobs. The effects of the crisis of the early 1970s, followed by economic restructurings that diminished demand for low-skilled labour, have been severe on Vaulx-en-Velin, much more than on other communes in the region.

Table 6.1 shows unemployment rates for selected years. In the 1970s, the level of unemployment in Vaulx-en-Velin was slightly higher than the levels of unemployment for Lyon, the department, the region and mainland France. In the 1980s, unemployment rates started to go up in France, and the Rhône-Alpes region in general was affected by job losses, although the

Table 6.1 Unemployment rates in Vaulx-en-Velin and its urban policy neighbourhoods compared to departmental, regional and national rates for selected years (%)

	Vaulx-en-Velin					
	Urban policy neighbourhoods	*Commune*	*Lyon*	*Rhône*	*Rhône-Alpes*	*Mainland France*
1975	–	4.52	3.58	3.15	3.04	3.82
1982	–	14.01	7.65	7.58	7.71	8.93
1990	17.9	16.0	9.25	8.79	8.97	10.88
1999	28.4	23.3	12.56	11.44	11.05	12.85

Sources: INSEE (1990); INSEE (1999a); INSEE-DIV (no date).

unemployment rate at the regional level has always been below the national level. Communes such as Vaulx-en-Velin were severely hit by diminishing industrial and manufacturing activities, due largely to their population characteristics (high proportion of workers, usually low-skilled). Compared to Lyon, to the department or to the region, the unemployment rate in Vaulx-en-Velin, from the 1980s onwards, has been twice as high, and unemployed people were concentrated mainly in the previous ZUP, which would later become one of the sites of urban policy. In 1999, almost a quarter of the active population of Vaulx-en-Velin, and more than a quarter of the active population of its social housing neighbourhoods, were unemployed. The rates were even higher for young people between 15 and 24 years of age (37.1%) and foreigners (29.9%).

The figures are even more dramatic for the social housing neighbourhoods of Vaulx-en-Velin that have been included in urban policy since 1984. They consist of the former ZUP (including Mas du Taureau, where the incidents had started) and La Grappinière. Table 6.2 offers a comparison between the commune of Vaulx-en-Velin and its urban policy neighbourhoods for 1990 and 1999.

Like many *banlieues* included in urban policy, Vaulx-en-Velin has a relatively high proportion of foreigners and young people, both of which suffer disproportionately from unemployment. We should note that unemployment rates have gone up in all of France since the 1970s, and not only in

Table 6.2 Unemployment rates and population characteristics of Vaulx-en-Velin and its urban policy neighbourhoods, 1990 and 1999 (%)

	Urban policy neighbourhoods of Vaulx-en-Velin		Vaulx-en-Velin		Mainland France	
	1990	1999	1990	1999	1990	1999
Unemployment	17.9	28.4	16.0	23.3	10.8	12.8
Unemployment (15–24 years old)	25.5	40.5	24.1	37.1	19.9	25.6
Unemployment (foreigners)	25.3	33.1	23.6	29.9	18.8	24.1
Less than 20 years of age	36.2	34.1	34.3	32.0	26.5	24.6
Foreigners	25.4	25.0	22.7	20.8	6.3	5.6

Source: INSEE-DIV (no date).

Table 6.3   Rates of increase in unemployment levels in Vaulx-en-Velin and its urban policy neighbourhoods compared to departmental, regional and national rates for selected periods (%)

	Vaulx-en-Velin					
	Urban policy neighbourhoods	*Commune*	*Lyon*	*Rhône*	*Rhône-Alpes*	*Mainland France*
1975–82	–	210	114	141	154	134
1982–90	–	14	21	16	16	22
1990–9	59	45	36	30	23	18

Sources: Calculations from INSEE (1990); INSEE (1999a); INSEE-DIV (no date).

banlieues like Vaulx-en-Velin. However, these (former) working-class areas were the most hard-hit ones following the crisis, where unemployment rates went up dramatically in the years to follow, as Table 6.3 exemplifies by the case of Vaulx-en-Velin.

These figures may be taken to indicate two issues that concern not only Vaulx-en-Velin, but social housing neighbourhoods in *banlieues* in general. First, despite having been included in urban policy programmes for years, such neighbourhoods suffer from an aggravating unemployment problem, leading to severe conditions in the worst affected neighbourhoods where one in four workers – almost one in two among young people – is out of a job. Second, as the above figures for selected periods show, the transformations of such neighbourhoods are closely linked to restructuring processes: that is, to dynamics that well exceed the perimeters of designated intervention areas of urban policy.

What, then, happened in Vaulx-en-Velin after the incidents of 1990, where, as the National Council of Cities report put it, a new relationship of citizenship would develop in the aftermath of the revolts?

A 'Thirst for Citizenship'

The *association Agora* has its origins in a committee created in the aftermath of the incidents following the death of Thomas Claudio in 1990, which was seen as 'the last straw' (interviews with Abdel Della and Pierre-Didier Tchétché-Apéa). Called '*comité Thomas Claudio*', this group had three objectives: following the juridical process about the controversial accident, keeping the public informed about the evolution of the process, and finding money to hire lawyers and continue the process.

Agora was created out of this committee in November 1991. The association's objectives included educational and cultural activities, and services for the inhabitants of the neighbourhood, ranging from preparation of CVs to help with legal questions, especially on issues related to immigration. The activities of the association were stated as the following in its statute:

- Denounce and fight against all forms of exclusion.
- Development and encouragement of associative life, leisure and sports activities.
- Information on the importance of practices of civic rights.
- Promoting the cultural expression of the inhabitants of the commune.
- Encouraging the creation of socio-economic activities.
- Contributing and encouraging the participation of young people to their city.
- Encouraging and supporting all such initiatives.[1]

In addition to these, urban policy in Vaulx-en-Velin would quickly become an occupation for *Agora*, with the objective of following, and questioning if necessary, the activities of the local government within the framework of this policy.

Before the incidents of 1990, there were large-scale renovation projects in Vaulx-en-Velin carried out under the auspices of urban policy. These projects focused mainly on improving the physical environment.[2] There were, however, problems that were not adequately addressed by such a focus. The incidents, in this sense, were influential in helping to highlight deeply embedded problems. They

> raise even more sharply issues that questions of architecture and urbanism could not deal with adequately. There are questions of justice, social justice, discrimination, of discrimination in every respect, for access to knowledge, for access to housing, for access to jobs . . . and beginning with a revolt which was based on the death of a young person of the neighbourhood, all these issues resurfaced subsequently. (Interview, Yves Mena)

Yves Mena used to work in the mayor's office, first on economic development, and later, after 1989, on establishing communication with the inhabitants of the neighbourhood. He resigned in 1995 and started working with *Agora*. Following the incidents, he was asked by the mayor to establish dialogue with the young people of the neighbourhood who were willing to participate in the life of their *cité*. The mayor, at first, was ready to respond to this '*thirst for citizenship*', as Mena puts it.

It was in this context that some of the members of *Agora* started to get involved with a social centre, *Le Calm*, that then existed in the neighbourhood of Mas du Taureau. The centre had never worked very well. However, following the incidents, some of the young inhabitants of the neighbourhood became more engaged politically, and used the centre as a setting. In the winter of 1992, its director resigned, and the administration of the centre was taken over by the young inhabitants of the neighbourhood, including the two founding members of *Agora*, Abdel Della and Pierre-Didier Tchétché-Apéa. Pierre-Didier is the president of *Agora*. He is also a founding member of the association *Divercité*, based in Lyon, which tries to bring together the associations of *banlieues* in the eastern Lyon region.

Despite organizational and financial problems, the social centre quickly acquired an important place in the neighbourhood. This, as Abdel and Pierre-Didier argued, was perceived as a possible threat by the municipality, undermining its power in the neighbourhoods. Funding for the centre stopped, which eventually led to its closing. Then one day in November 1994, early in the morning, the centre was razed to the ground upon the order of the mayor.

The destruction of the social centre put an end to efforts to establish dialogue between the municipality and the young inhabitants of the neighbourhood. From the latter's perspective, the mayor's initial enthusiasm following the incidents had passed, and his perspective on the population of the city had changed remarkably; they were 'too poor, too foreign origin, always too something' (interview, Yves Mena). New notions emerged on the municipality's agenda. 'Social mixity' (see Chapter 1) was one of them, which was first introduced by a 1991 law in the framework of urban policy – the so-called LOV, as we have seen in Chapter 4.

Thus started the construction of a new city centre with an attempt to 'change the image' of Vaulx-en-Velin.

A Toil of Two Cities (in One)

The new centre of Vaulx-en-Velin was originally supposed to be at Mas du Taureau, where heavy investments were made before the incidents. After the incidents, however, it was decided that another new centre was necessary. Mas du Taureau was abandoned, so to speak, and with the money pouring in following the incidents, the construction of a new centre started from scratch. The location designated for the new centre was away from the social housing stock, to its south-east, on the other side of the fairly wide and busy D-55 road. When Jean Pirot first arrived in Vaulx-en-Velin in 1994, he was working for the *Conseil Général* (departmental council in France). Since 2001, he has held a position as project manager (*chef de projet*), and is responsible for social and associative life in Vaulx-en-Velin.

He recalls the location designated for the new centre; it was a vacant area where there were no residential buildings – where there was, indeed, nothing at all:

> Say, ten years ago, not even ten, when I arrived, the effect was strange: there was the city hall there, and in front, there, it was wasteland, grass, a field. . . . (Interview, Jean Pirot)

The municipality's decision to move the centre away from the social housing neighbourhoods, just like its decision to close the *Le Calm* social centre, had immense symbolic value. The decision to build a new centre away from Mas du Taureau gave an impression of abandonment, and the municipality's attempt to bring in 'better' inhabitants in the name of 'social mixity' added insult to injury. 'Social mixity' was now the municipality's policy principle, introduced and institutionalized by urban policy, but it was seen as an offensive statement by those 'on the other side',

> as if people here were of inferior quality, and that to improve them, a greater quality of people had to be brought in from elsewhere, who knows where from, actually. (Interview, Yves Mena)

> [I]n fact they, they wanted to create a new centre, with new inhabitants, in order to bring in new people. (Interview, Abdel Della)

This was how the attempt behind the creation of a new centre was interpreted by the young people in the social housing neighbourhoods, an interpretation that is not contested by the municipality:

> The choice that was made to create a city centre, it's a strategic choice of the mayor, which was to say: if we don't make *a proper city*, with a real city centre, we'll never be able to attract *a different population*, and notably what we call middle classes, which can contribute to social mixity, and from there, precisely, we can reverse *the ghetto phenomenon*. If we don't have population mixity, *the ghetto will remain*. [. . .] It's true that by making this decision, the mayor indeed took the risk of not spending the money to develop cheaper local equipment in the neighbourhood, that's true. (Interview, Jean Pirot)

But why would Vaulx-en-Velin, characterized by a fairly rich diversity of population, have ghettos? The project director explains this as follows:

> The problem is that today, that the ghetto has become *communitarian*. And there, we have another problem. That is, in fact we weren't in, that is, the word ghetto is always difficult to use . . . when one lived in a *grand ensemble*, in a ZUP, was it a ghetto up until the 1980s? No, it only became one from

the moment when a certain type of population, in particular *with community forms*, moved in. (Interview, Bernard Brun)

Bernard Brun has been working in Vaulx-en-Velin since 2000, and he is a project director for the *Grand Projet de Ville* (GPV) of Vaulx-en-Velin, within the framework of which a new city centre, much to the discontent of *Agora* members, is being constructed. The 'certain type of population' Brun has in mind, given the time of the transformation he is talking about, is likely to be non-European immigrants, for it was around that period that they started gaining more access, and becoming more visible, in social housing neighbourhoods. 'Communitarian', in this sense, implies non-European immigrants, and Brun's remarks exemplify the common usage and connotations of the term in debates around the *banlieues*. In contemporary France, the terms 'ghetto' and 'communitarian' invoke a very precise population: non-European immigrants. As we have seen in Part II, these terms have become part of state's statements on the *banlieues* from the 1990s onwards, constituting them as incompatible with the ideal of 'one and indivisible' French republic.

But let us go back to the new city centre debate in Vaulx-en-Velin. Although the decision to build the new city centre away from the social housing neighbourhoods in itself caused major discontent, what made it even more contentious was that it followed a similar pattern in terms of investment strategy and decision-making in Vaulx-en-Velin. For the members of *Agora*, the actions of the municipality in the framework of urban policy have been directed less towards the population already living there, and more towards the 'development of the city' in order to attract another, 'better' type of population. The effects of urban policy, from their point of view, have been hard to discern:

Are you familiar with urban policy?
Yeah, more or less, yeah.
Well, how do things work in this neighbourhood? This is 'difficult neighbourhood' [one of the names for designated urban policy neighbourhoods], isn't it?
Yeah, this is . . . yeah, I don't really get it, I don't understand what they want, I don't understand, I don't see the point.
What's wrong?
[*silence*] What's wrong?
Because you know the problems of the neighbourhood and all that.
[*silence*] I couldn't . . . I don't see, I don't see . . . what urban policy is, what the point is, why . . . what the use is, what it could improve . . . It's been ten years, it hasn't changed, so I don't see what use it could be, I don't know . . . apart from the money they hand out from up there, and that we, at least us in the *Agora* association, we never see any of it, or only a tiny portion, I don't know where the rest goes . . . I don't get it.
OK.

I . . . urban policy is nothing but words!
This neighbourhood has been included in urban policy for twenty years . . .
Yeah, so, there's no change, there's no, there's not a thing that there's nothing, it's the same thing . . . I've lived here since '74, I don't see a difference.
But there's money pouring in . . .
And yet, they tell us there's lots of money allocated to the neighbourhoods, but there's nothing, there's no change. (Interview, Abdel Della)

A similar critique of urban policy is made by Lahlou, one of the founding members of the MIB (*Mouvement de l'Immigration et des Banlieues*) first encountered in Chapter 4:

> First, urban policy has never been very transparent, you see. It has never been very transparent; we had asked for audits. [. . .] Whereas if you ask the inhabitants of *banlieues*, they will tell you, yes but us, we don't know what has happened. In our daily life, nothing has changed; politically, nothing has changed. [. . .] For 20 years now there has been urban policy, you see, since the arrival of Mitterrand basically. I don't think it has changed much for the inhabitants, and it's even getting worse. Misery, despair . . . (Interview, Lahlou, MIB)

Investment strategy, for the members of *Agora*, is one source of discontent. Another one relates to how decisions are made. As they argue, even in 1990, when Vaulx-en-Velin was presented as an exemplary neighbourhood of urban policy's renovation programmes, realized projects were far from the expectations of many inhabitants. They reflected more the 'personal desires' of the local officials than the expectations of inhabitants; they 'were not constructed with the inhabitants' (interview, Abdel Della); 'the inhabitants do not exist' (interview, Pierre-Didier Tchétché-Apéa). Either the decisions are made and then applied, or meetings are held in the name of 'participation of inhabitants' where already conceived projects are presented to the public – 'participation simulacra', as Pierre-Didier calls them (Tchétché-Apéa, 2000: 83).

> Because the decisions that are made about these neighbourhoods, they are made, there's no public debate about urban policy. Before setting the money on urban policy, do they, is there, I don't know, when they decide on agricultural policy, or when they want to reform Social Security, it's done with the actors, both political and from civil society. While, as far as we're concerned, on urban policy, there's no public debate! (Interview, Pierre-Didier Tchétché-Apéa)

Although the members of *Agora* are quite dissatisfied with this form of decision- making, the mayor of Vaulx-en-Velin seems to be satisfied with the process, and has now more projects in mind for the social housing neighbourhoods:

> [T]he centre of Mas du Taureau, which has, by the way, been taken care of today, *with the state, with the urban community* [a grouping of communes in an urban area], we even deliberated *in the municipal council* a few months ago, we started a survey, which I shall present to the inhabitants in the autumn, we started a study to start, to initiate a process of urban renewal on the area of Mas du Taureau and the commercial centre of Mas du Taureau.
> *OK.*
> So, I think within two years, the time to set up operations and *to concert it with the inhabitants*, operations will be started, and they'll be of considerable importance. (Interview, Maurice Charrier)

Maurice Charrier has been the mayor of Vaulx-en-Velin since 1985. He was the deputy mayor before, and was responsible for urban planning. He was a member of the Communist Party, but he no longer is, although he considers himself 'still from the communist culture'. Although he is not opposed to the participation of inhabitants, there seems to be a major disagreement between him and the members of *Agora* as to what this should entail.

> I am for the participation of inhabitants, but not for demagogy. Let's get this straight: there are things to be discussed with inhabitants, and others that are not discussed with inhabitants. (Interview, Maurice Charrier)

> And in this respect I think Maurice Charrier wants there to be, he always says: there are negotiable things, and non-negotiable ones. So today, the role of politicians is to say: this is non-negotiable; that, however, can be negotiated. (Interview, Bernard Brun)

The discord resulting from different ways of conceiving 'participation' surfaced when *Agora* opposed the mayor's project for the new city centre and the proposed construction of the planetarium as part of that centre. For the mayor, the new centre will establish Vaulx-en-Velin's position in the region as a pole of attraction with such facilities as the planetarium and cultural centre. His effort seems to be to change the image of this notorious *banlieue*. There is nothing extraordinary, or inherently bad, in this effort, which reflects entrepreneurial urban development strategies. Such strategies, however, do not effectively respond to the needs and expectations of the inhabitants of dispossessed neighbourhoods, which also seems to be the case in Vaulx-en-Velin, where the mayor's development strategies and the ways in which they have been put into practice aggravated the tensions between the municipality and *Agora*.

Figure 6.1   Social housing neighbourhoods of Vaulx-en-Velin seen from the new centre (*Source*: Photo by the author)

Agora opposed the planetarium project, and demanded, instead, the construction of social centres in the neighbourhoods ('*maison de quartier*'). The mayor was in favour at first, and a convention was signed with the region in 1991 in the framework of urban policy, providing Vaulx-en-Velin with a large amount of money to finance the construction of three social centres in its social housing neighbourhoods, including Mas du Taureau. Later, however, the mayor changed his mind, and opted for more spectacular projects, including the planetarium. As the members of *Agora* claim, there had been a 'misappropriation' (*détournement*) of funds; the money that was originally raised for the construction of three social centres was used to finance the construction of the planetarium in the new city centre.[3] The social centres, eventually, were not constructed. What resulted was a divided city, with the social housing neighbourhoods – with their existing inhabitants – and the new centre – with its expected new inhabitants – on either side of the D-55 road (Figures 6.1 and 6.2).

> There you have the town hall, the sports complex, the school, a professional high school [. . .], there you have the planetarium, you have the youth service which is there, you have the cultural centre which is there, you have . . . in fact, you have the city centre which is nearby, you have the post office, everything is centred over there. (Interview, Abdel Della)

Figure 6.2   The new centre of Vaulx-en-Velin seen from the social housing neighbourhoods (*Source:* Photo by the author)

Or, as Pierre-Didier puts it: 'We're really in, in worlds out of sync [*mondes décalés*].'

Whose List is More 'Communitarian'?

The year 1994 was a turning point for *Agora*, for two reasons. First, with the tensions arising first from the closing of the *Le Calm* social centre, and then from the construction of planetarium and new city centre, its relation with the municipality came to an end, which meant that it no longer benefited from financial aid distributed through the municipality. Second, its members started to question the limits of purely associative actions, which eventually led to a decision to enter the political arena through a more conventional route. The political party *Le Choix vaudais* was thus created, and a list was prepared for the 1995 municipal elections with the involvement of *Agora* members.

Le Choix vaudais started its campaign only a couple of months before the elections, and its electoral list was prepared with a stated aim of representing the inhabitants of Vaulx-en-Velin. 'It [i.e. the list] made us *enter a public space*. It made us *visible* and carried new demands and claims' (Tché-tché-Apéa, 2000: 82; emphasis added). But the public space seemed unable to accommodate any new demands and claims, especially if they came from the 'certain type of population' Brun was talking about. The members of the party were accused of presenting an 'ethnic/communitarian list' despite the fact that all the members presented on the party's list were French citizens on paper, as the rules require. This accusation, however, did not come from the extreme right:

> And this list, by the left, because the left was the harshest, the most violent against us, this list was called communitarian list, ethnic list. [. . .] And us, those attacks, they were permanent, violent, really, really intolerable. (Interview, Pierre-Didier Tchétché-Apéa)[4]

The counter-argument was that it was 'their' own list that was 'communitarian'. When electoral lists consist of the 'real' French and perhaps a few so-called *Arabes de service*,[5] they are not seen as 'ethnic' or 'communitarian'.

> Whose list is ethnic? Yours! Because your list, it's of a single colour, it's ethnic! Ours is, there are several, it's *the citizens of Vaulx-en-Velin*! That's what they are, and that's what we want to represent. You see? [. . .] There isn't, there's no movement in France that comes from the [social housing] neighbourhoods or from immigration that claims a communitarian project! There's none! If you know of one, tell me! There isn't any! Even during the March, the March in '83 [the March for Equality and Against Racism – see Chapter 3], it was about equality! White, black, *beur*, all that. It was, it is about equality. (Interview, Pierre-Didier Tchétché-Apéa)

Interviews with the members of *Agora* suggested a mistrust towards 'the left', mainly the Socialists and the Communists, who traditionally had a strong presence in the social housing neighbourhoods of *banlieues* – which, indeed, have been their strongholds. The remarks of *Agora* members, however, suggest that this may no longer be the case.

> A youngster from this city, how could you tell him or make him believe that the left was better, fairer, more egalitarian, more humanistic, and cared for his future, while he lived in a precarious situation, in housing, unbearable housing, in a, and in schooling also a victim of segregation. [. . .] And since then, the dispute has just got worse. And we see today, when there are elections, we see neighbourhoods like these are no longer strongholds of the left systematically, automatically. [. . .] So lots of people don't vote, but among those who do, well, it's no longer an automatic vote in favour of the left, at least for this left, which is not going to bring the solutions it promised. (Interview, Yves Mena)

'This Left which is not going to bring the solutions it promised' – that was a recurrent theme in all the interviews with the members of *Agora*. The Socialists are accused of 'playing with' voting rights for immigrants, which was first promised by Mitterrand in 1981. This promise, Abdel argues, served the left well in the 1980s for getting the votes of second-generation immigrants. Now, however, the left not only has lost its credibility, but has even become an 'enemy':

> Yes, I think it [voting rights for immigrants] was useful to them during the '80s. But whatever, now people in the neighbourhood no longer vote, or vote very little. For them, left and right, the National Front, all that is the same shit. It's the same thing . . . and they're all alike. There's no difference. And at a pinch, they prefer to have someone from the National Front in front of them, rather than someone who lulls them to sleep every day, and in the end has nothing for them. They prefer someone from the National Front. At least with the National Front, things are clear. There you are, it's, it's better. It's better like that. (Interview, Abdel Della)[6]

Similar remarks were made by other young inhabitants of the neighbourhood, with whom I talked but could not interview with a tape recorder.[7] This antagonism towards the left does not follow merely from unfulfilled promises and aggravating conditions in the *banlieues*. It also has to do, as the interviews suggest, with the paternalistic, even 'colonial', way in which such neighbourhoods are governed under the left:

> And what's more, as these cities are usually run by the left, the left always considered immigration could only vote for the left, so we were hostages of the left, which didn't understand we want our autonomy. (Interview, Pierre-Didier Tchétché-Apéa)

Abdel, from Algeria, argues that the mayor 'wants what they did in other countries .. the colonizers, and that's what they want'. Pierre-Didier, from the Ivory Coast, talks about 'a sort of colonial management of these neighbourhoods'. There was, however, no reference to colonization in the interview with Yves Mena, whose parents immigrated from Spain. What this suggests is that there are deep injuries that derive from France's colonial history. Abdel puts it rather bluntly: 'The Algerian war is not over in France.' Indeed, a recently organized movement put this issue at the forefront of their claims with a call, diffused in January 2005, with the title 'We are the natives of the Republic!', which argued that France had been – and still was – a colonial state:

> Quite independently from their actual origins, people of the 'neighbourhoods' are 'indigenized', relegated to the margins of society. The *banlieues* are said

to be 'lawless' ['*zones de non-droit*'], and the republic is called upon to 're-
conquer' them. Arbitrary identity checks, a variety of provocations, persecu-
tions of all kinds are rife, while police brutality, sometimes of the most
extreme form, is seldom punished by a judicial system which has double
standards.[8]

But let us go back to Vaulx-en-Velin. Both *Agora*'s political activities
and the attempt to enter the political arena through *Le Choix vaudais* were,
as Pierre-Didier puts it, attempts aimed at a certain political autonomy in
order not to be 'managed as though we were kids! In a colonial way, in a
paternalistic way!'

> [T]he act consists in . . . that's citizenship for us. [. . .] So, what do you want,
> you want actors, but when people become actors, it's a problem for you?
> No, what you want is to go thinking in people's place, just go on thinking
> in our place, representing us. Well, that's over! (Interview, Pierre-Didier
> Tchétché-Apéa)

So no salvation from 'the Socialist parties, who claim to be representing
people, or the Communist Party, which claims to be representing people'
(interview, Abdel Della). What about, then, nationally renowned associations that were created in the mid-1980s, the heydays of immigrant activism? The interviews suggest that a similar distrust exists towards such
associations, of which *SOS-Racisme* and *France-Plus* are the most eminent
ones, for diverting locally formed claims. A similar observation was made
by Chabanet concerning these associations. *France-Plus* and *SOS-Racisme*,
he argued, 'no longer have any credibility at the local level' (1999: 357;
see also Begag, 1990). The distrust towards such associations has to do
with the nature of their activities and with their strong links with major
political parties. The Socialist and Communist Parties, for example, are
accused of supporting associations that emerge from their own local organizations rather than leaving the initiative to the inhabitants of the neighbourhoods themselves. For the members of *Agora*, these associations
appropriate claims formulated at the local level without posing serious
challenges to the established order of things.

> When they created *SOS-Racisme*, they created *SOS-Racisme* to avert the *beur*
> movement from their claims and the autonomy it was aiming at. So they
> created *SOS*, they oriented the fight and the struggle on a struggle that means
> nothing, anti-racism, which finally boils down to actions to have people
> allowed into night-clubs, it's bullshit! So our future is night-clubs? So, the
> republic, for us, is to mean getting into night-clubs? It's just incredible! And
> at a time when you're organizing concerts, you have 100,000 people who
> come to Paris, you just say, you give speeches, but it's not in speeches that

you change, that you change things! [. . .] When you have 100,000 people, you are in a position to make society evolve! Politically! Not in the way, meeting to eat couscous, to dance Arabic dances, folklore, and there you are, everybody go home, and it's over, no! (Interview, Pierre-Didier Tchétché-Apéa)[9]

With these reservations about the left and nationally renowned anti-racist associations, *Agora*, supported by other associations in Vaulx-en-Velin, engaged in the campaign for the 1995 municipal elections. After two months of campaigning, *Le Choix vaudais* obtained only 7.2% of the votes, and did not qualify for the second round. The abstention rate was above 40% in both rounds (44.7% for the first round, 41.6% for the second). Although Maurice Charrier still remained the mayor, the National Front became the second major political force in the commune (31.0% of the votes in the first round, 33.5% in the second), following a campaign organized around the slogan 'Vaulx-en-Velin, a French city, it is possible!'[10]

The two major campaign themes of *Le Choix vaudais* were the participation of inhabitants and the improvement of living conditions. Although the initial enthusiasm now seems to have weakened, *Agora* members still have faith in the party, which, for them, represents 'the citizens of Vaulx-en-Velin'. The way *Agora* conceives citizenship, however, goes beyond a formal status, and involves practices through which inhabitants stake claims and engage in political actions in matters concerning their everyday lives.

> Citizenship, for us, goes well beyond a strict juridical definition that boils down to a national status, to a capacity to vote in elections. Our citizenship consists in being consulted on local projects that directly affect our everyday lives. (Agora, 1995: non-paginated)

> The concept of *Agora* is essentially the *appropriation of lived space* by reflection and by construction of ideas, and their materialization through the initiative of inhabitants. The difficult context in which we live demonstrates how fundamental it is to encourage the involvement of citizens, of inhabitants in the *life of the Cité*. It is more necessary than ever to take concrete and effective action to make inhabitants aware of their responsibilities in the neighbourhood. (Document Agora, no date; cited in Chabanet, 1999: 354–5; emphasis added)

Le Choix vaudais, in this sense, was seen as a struggle for developing a form of local citizenship: 'In this citizens' struggle', Yves Mena insists, 'we're *citizens of Vaulx*, and that's it'. Pierre-Didier makes a similar remark:

> So, we think, OK: in the discourse of the republic, all that, it's: make citizenship, be citizens, etc. When you become a citizen, you see it poses a problem,

it gets in their way. And this action was, in '94 too when funding was suspended, saying, right, we must invest in the political field. So, when we invest
in the political field, us, as French citizens, because you have to be French
to form a list, you make a list *related to your city*, and that was really very very
important. [. . .] And this list [. . .] was based on a will to represent all *the
citizens of Vaulx-en-Velin*. (Interview, Pierre-Didier Tchétché-Apéa)

The engagement of *Agora* in the life of the *Cité*, their urban definition of
citizenship, and *Le Choix vaudais* experience are telling in that, on the one
hand, there is an attempt to constitute political subjectivities detached from
ethnic origins, and linked to local belonging, to the city and its lived spaces.
On the other hand, there is a counter-effort to repress this form of political
subjectivity by confining it to 'ethnic origins' through accusations of 'communitarianism', unacceptable under the republic. There is, in other words,
an attempt to open up political spaces in a context where the space of the
political seems well delimited.

Conclusions: Acting on the Spaces of the Police

The post-war trajectory of Vaulx-en-Velin gives an insight into the characteristics of social housing neighbourhoods under urban policy. The account
presented here is suggestive in three respects: everyday problems of *banlieue*
inhabitants, youth in particular; local reverberations of urban policy; and
political mobilizations in *banlieues* and the limits placed upon them.

One of the immediate repercussions of the incidents of 1990 was the
engagement of the French Intelligence Service with the question of *banlieues*, with a stated aim to fight 'urban violence' – a notion that entered
policy discourse following the incidents. Interviews with *Agora* members,
however, suggest that there are other forms of violence experienced – not
inflicted – by the *banlieue* youth on an everyday basis, notably in the job
market and in relations with the police. Concerning the former, Pierre-
Didier tells a familiar story: you see a job opening, give a call and everything
seems to go fine since your voice does not reflect your skin colour, but
when they see you, they tell you the job has already been taken. 'You think
yourself as normal, like the others, with rights, with respect, etc. This is
very violent, when you are rejected like that, this is very, very violent!'

It is, however, not only the skin colour; living in a highly stigmatized
banlieue like Vaulx-en-Velin aggravates the situation. There are, for example,
strategies for applying for a job, such as giving an address in Lyon or the
neighbouring Villeurbanne, and not in Vaulx-en-Velin. During the interviews, Jean Pirot stated that the North African population in the commune
was particularly discriminated against, and provided a telling, if fictive,

example about the negative effects of territorial stigmatization of *banlieues* like Vaulx-en-Velin:

> A young person from Vaulx-en-Velin, if he's called Mohammed, he has a dark complexion, and he says he lives in Vaulx-en-Velin, he can have a postgraduate qualification, if he applies for a job in some part of western Lyon, he won't be hired. If he has, if he's called Mohammed, he's not too brown, and he says he lives in, say, Oullins, his chances are better. (Interview, Jean Pirot)

Strong territorial stigmatization also affects relations with the police. Discriminatory practices of the police against *banlieue* youth (a common form of which is constant identity checks) and the hostile relationship between the police and *banlieue* youth have been observed by many researchers (for the case of Vaulx-en-Velin, see Begag, 1990; Chabanet, 1999; Essassi, 1992). In Vaulx-en-Velin, the relationship between the youth and the police has always been full of tension, even before the incidents of 1990. As Abdel put it, being a 'young person of a neighbourhood, being a second-generation immigrant, being an Arab, having a darker complexion' pose particular problems in relations with the police.

This strong territorial stigmatisation, as we have seen in Part II, is not unrelated to the official framing of *banlieues* – through urban policy as well as other state statements – in increasingly negative terms. Vaulx-en-Velin is also indicative in that it shows the effects of urban policy measures at the local level. Local reverberations of urban policy vary, to be sure. But the Vaulx-en-Velin story shows at least how fuzzy policy objectives such as 'social mixity' are open to a number of interpretations – which, in this case, resulted in the construction of another city centre away from social housing neighbourhoods with the aim of attracting 'better' populations. This suggests that there may be unchecked discrepancies between the stated aims of urban policy and their local interpretations, and that such discrepancies may translate into gaps between investments and the expectations of inhabitants. Discontent arising from such gaps, combined with mass unemployment, negative effects of territorial stigmatization, deep injuries deriving from France's colonial history, and a sense of political exclusion, aggravate tensions in *banlieues*. The Vaulx-en-Velin story is telling in this sense as it points to some deep-rooted problems facing *banlieue* inhabitants, youth in particular. The persistence of such problems suggests that it would be misleading to interpret *banlieue* revolts merely as intrinsic acts of violence; they are spontaneous responses, if violent at times, to such problems – a point that I will try to make more strongly in the following chapter. The incidents of 1990, for example, were neither pre-conceived nor organized; the young people simply revolted, as Abdel put it.

But to my mind, perhaps the most significant implication of this account is that *banlieues* also become sites of political mobilization – or of 'insurgent citizenship' – that introduce new political identities into the city (Holston, 1998), with democratic ideals despite their constitution increasingly as 'badlands' through state's statements. As the story of *Agora*'s political engagement implies, however, such mobilizations encounter problems, which mainly arise from two sources: the 'workings of the system' and the discursive framing of *banlieues*. Workings of the system here refers to the way in which associations, like *Agora*, are funded and allowed to participate in decision-making processes within the framework of urban policy. Urban policy devotes money for funding the activities of associations in selected neighbourhoods, but this money goes through the departmental council and the municipality – through, in other words, the established institutions of 'the police'. The municipality, then, has the possibility of creating a context of competition between different associations demanding funding for their activities. This scheme has two adverse effects for democratic politics. First, it makes it possible for the municipality to stop funding associations whose activities contest the *status quo*. The case of *Agora* is exemplary here, which was seen as too involved with the political affairs of the city, leading eventually to the suppression of its funding. Had *Agora* members focused their activities on issues such as music, sports, and so on, they would not have had difficulty in receiving funding. Getting involved 'too much' in the political life of their city, however, did not quite fit into the established framework of associations. When asked about this issue, Jean Pirot, who is now responsible for social and associative life in Vaulx-en-Velin, responded as follows:

> I understand your question, and it's difficult, the criticism that can be made, and you hear it here and there, is: the dice are loaded with urban policy, because you pretend you're involving the citizens, and taking measures for them to be more like citizens, and a bit more responsible, and when they are, and they claim a part in the political choices that are made, then you say: stop! It's true, it's a real difficulty, that's true.
> *So practising sports, organizing parties, things like that are more . . .*
> Yes, yes, roughly speaking, practise sports, organise meals, make music, some festivals, but whatever you do, don't get involved in . . . there you are, I know it's. . . . (Interview, Jean Pirot)

Or, as Nordine puts it:

> There were associations in those neighbourhoods, they could have been funded, been given means to implement policies in response to the problems of these neighbourhoods. And what did they do? Well, they were emptied out, corrupted. [. . .] You have to ask for subsidies, for funding, but what

will they give you, nothing. If you are not . . . if you are not sheep-like . . . I don't know . . . you have to agree with the way they see things or else, if you criticize them, they have the means to 'ruin' you. (Interview, Nordine, MIB)

The second problem engendered by this scheme is that it impedes the possibility of united – therefore potentially more effective – actions by associations as it puts them in competition for limited resources. Thus, it leaves little or no room for manoeuvre for associations that are politically engaged (Chabanet, 1999; Donzelot and Estèbe, 1999; Nicholls, 2003), while leaving intact the concentration of power in the hands of the few. As Hunt and Chandler (1993: 71) wrote:

A more pertinent critique of the French system is that, for all its localism and flexibility of practice, the structure is not particularly democratic. While it allows citizens access to the decision-making system so that elites may circulate, it places power largely in the hands of small local elites who govern their territories through local and national influence as, in some cases, personal fiefdoms. It is a system much more based on paternal central direction, and hence far less open to interest group involvement or popular pressure.

Given this rigid institutional structure, it is very difficult to constitute structures of counter-power. This also has to do with the decentralization reforms of the early 1980s, which increased the powers of mayors while those of inhabitants remained rather weak (Chabanet, 1999). Consequently, mayors enjoy a considerable amount of power at the local level with little or no possibility for contestation.

At the same time, by so doing, it [decentralization] makes mayors real, real overlords, in a way, on their estates, and it reinforces their power as elected officials, you see, as elected, as first magistrates. So that's, at the same time, it's positive, but it depends how it's used. (Interview, Pierre-Didier Tchétché-Apéa)

In Vaulx-en-Velin, no association has adequate political influence to affect decisions (Chabanet, 1999; Essassi, 1992). The space of political possibilities is limited, as Yves Mena argues:

So urban policy, roughly, it means policy for the *banlieues*, doesn't it. [. . .] They say OK, we'll call it urban policy [*la politique de la Ville*], but in fact it applies to *banlieues*. [. . .] Without a real intention to invest massively in education, in employment, in local democracy, in the participation of inhabitants . . . the intention is not there. The will of power, either local or governmental or regional, the will of power is always to control and master, what's done everywhere. *And not to leave space, or to leave as little space as possible*, to controversy, either political or social, or anyway, to political debate. (Interview, Yves Mena)

The account presented here, however, suggests that there are attempts to open up political spaces in *banlieues* by acting on the well-delimited and over-determined spaces of 'the police'. Both *Agora* and *Le Choix vaudais* can be seen as local political formations with a decidedly urban and spatial focus, using space as an organizing principle in the formation of political claims. This was most clearly expressed in the urban-spatial definition of citizenship and the 'appropriation of lived space' emphasized by *Agora* and *Le Choix vaudais*. But one of the difficulties such formations encounter, in addition to the workings of the system, is the discursive framing of *banlieues*, which was evidenced by the 'communitarianism', or 'ethnic separatism', accusation. Such accusations, to be sure, are in no way unique to this case; deplorable as they are, they are all part of the local political game. But they are not unique to Vaulx-en-Velin either; the same kind of accusations has been used to undermine similar political mobilizations. As Lahlou of the MIB puts it, whenever they tried to 'have a say', they were accused of 'communitarianism'. He cites a similar incident that had occurred in Ivry (a *banlieue* to the south-east of Paris), where letters were sent to the inhabitants stating that behind the mobilization of the youth were Muslim fundamentalists. What is important here is that the accusations used to undermine such formations and to legitimize controversial local projects (in the case of Vaulx-en-Velin) are all part of the state's statements. They are all institutionalized by urban policy, given a sense of legitimacy, made part of the police order. The Vaulx-en-Velin story, in this sense, shows the local reverberations of the consolidation of a police order and the implications of the discursive framing of *banlieues* through state's statements – what I have referred to, in Chapter 2, as urban policy as a place-making practice, affecting the everyday lives of people.

And what about the remarks of the National Council of Cities that were presented at the outset of this chapter, urging that Vaulx-en-Velin become a reference for 'a new relationship of citizenship'? '[T]hat discourse, you see, about citizenship, on the implication of inhabitants, on responsibility, both individual and collective, all that is merely discourse' (interview, Pierre-Didier Tchétché-Apéa).

7

Voices into Noises: Revolts as Unarticulated Justice Movements

Why did they happen? This question was remarkably absent in the aftermath of the autumn 2005 revolts in the French *banlieues*. For many activists, social workers and researchers, the relevant question was why such revolts have not occurred more often given the state of many social housing neighbourhoods. Having done practically nothing to alleviate inequalities, prevent discriminatory practices and police violence – disproportionately felt by *banlieue* inhabitants, youth in particular – the repressive government set up by Chirac was more surprised by the magnitude and persistence of revolts than by the fact that they happened at all.

Like previous revolts, those of autumn 2005 were triggered by the deaths of young inhabitants, in which the police, once again, were implicated. Like previous revolts, they were spontaneous – not organized – uprisings.[1] Also like previous revolts, they took place in the social housing neighbourhoods of *banlieues*, practically all of which were among the priority neighbourhoods of urban policy. Unlike previous revolts, however, they were suppressed by exceptionally repressive measures by the French state. They revealed not only once again the geographical dimension of inequalities, discrimination and police violence, but also the contemporary transformations of the French state along increasingly authoritarian and exclusionary lines. In this chapter, I expand these similarities and differences by putting the 2005 revolts in context, and comparing them with the previous revolts of the last two decades with a focus on their geographies, triggering incidents and (obscured) political significance in the consolidated police order.

Before moving on, however, the subtitle has to be accounted for, since this chapter is not about 'justice movements' – organized or in the making – as such. It is about the nature of revolts in the *banlieues* of French cities since the 1990s and the responses of the French state to them. Such incidents are not covered in the literature on new social movements in France (see, for example, Appleton, 1999; Waters, 1998). They are not social movements in the more conventional sense either, if we follow Buechler's definition of social movements as 'intentional, collective efforts to transform social order' (2000: 213). They are neither pre-conceived nor organized, and they are not articulated as collective efforts aimed at transforming the established order. However, as I will try to show, they are not intrinsic acts of violence either. They all mobilize with a demand for justice and as reactions against perceived injustices. 'Let justice be done' or '*J'ai la haine*',[2] as was heard – again – during the revolts of autumn 2005. Unarticulated as they are, such incidents are nevertheless episodic mobilizations that manifest contention and raise certain claims. Yet, their political significance has been obscured by state-led articulations of *banlieues*, and the French state has increasingly interpreted and responded to them merely as acts of violence. Hence the title of this chapter – *voices into noises*.

Revolting Geographies

On 27 October 2005 three young men in Clichy-sous-Bois, a *banlieue* to the north-east of Paris, took refuge in an electricity substation in order to escape identity checks by the police – a form of daily harassment not uncommon in the *banlieues* towards youths, especially if they have a dark complexion. Two of them were electrocuted and one was seriously wounded. That the police actually chased them was officially denied, although the surviving young man stated the contrary. This was the triggering incident for the revolts, which started on 28 October in Clichy-sous-Bois, and quickly spread to other social housing neighbourhoods of 274 communes, lasting for two weeks. More than 10,000 vehicles were set alight, and more than 3,000 people were placed under police custody, of which one third were indicted.

Similar incidents had occurred in the *banlieues* as early as the 1970s, though, compared to them, the revolts of 2005 were unprecedented in terms of their magnitude and geographical extent. As we have seen, two major series of revolts had been most influential in shaping political debate around *banlieues*: the so-called 'hot summer' of 1981 and the revolts in Vaulx-en-Velin in 1990. The 1980s witnessed five large-scale revolts in the *banlieues*. The 1990s, on the other hand, saw 48 large-scale revolts, in addition to some 300 on a smaller scale, referred to as 'mini-riots'. On Bui-Trong's (1993) 'scale of indicators of violence in sensitive neighbourhoods', the large-scale revolts of the 1990s were of 'degree 8' – the highest on the scale. This meant that they were characterized by the presence of up to

200 young people of the neighbourhood in question, motivated by a sense of solidarity against institutions of authority (police, municipality, and so on), lasting four to five consecutive days, and with confrontations with the police. The 'mini-riots' of the 1990s, on the other hand, were of 'degree 7', which meant that they were marked by 'massive vandalism' and Molotov cocktails with the participation of three to 30 young people, and with no confrontation with the police (Bui-Trong, 2000, 2003: 15–17).

With a few exceptions, all the large-scale revolts of the 1990s shared two common features in terms of their geographies.[3] First, all but two of the areas where revolts occurred were priority neighbourhoods of urban policy. Out of the social housing neighbourhoods of 38 communes where such incidents occurred,[4] four had been included since the policy's inception in 1982, three had been included since 1983, 13 since 1984, and another 13 since 1989. All of these priority neighbourhoods experienced revolts following the so-called 'return of the state' in the early 1990s. Three of them were included in 1996, after having experienced revolts, and two of them have never been priority neighbourhoods.[5]

Second, all the large-scale revolts of the 1990s took place in social housing neighbourhoods, nearly all of them in *banlieues*. These neighbourhoods and the communes where they are located followed a similar pattern in terms of constantly increasing levels of unemployment following the economic crisis of the 1970s and the ensuing processes of economic restructuring. In 1975, unemployment rates in the communes where these neighbourhoods are located were about the same as the national unemployment rate (except in Toulon and La Seyne-sur-Mer, where it was close to 8%, twice the national rate). After that, all of these communes were severely hit by diminishing industrial and manufacturing activities. To give an example, in Mantes-la-Jolie in the Paris region (included in urban policy since its inception, experienced revolts in 1991), the unemployment rate went from 3.9% in 1975 to 10.3% in 1982, then to 12.1% in 1990, and 20.2% in 1999. In its social housing neighbourhoods where the revolts occurred (Le Val-Fourré), it went from 15.7% in 1990 to 25.7% in 1999 (INSEE, 1990, 1999a; INSEE-DIV, no date).

What these figures suggest is that there is an embedded unemployment problem, constantly aggravating and hitting, more severely than any other place, the priority neighbourhoods of urban policy in the *banlieues*, which were once working-class neighbourhoods with low levels of unemployment. Furthermore, as we have seen in the previous chapter, the spatial designation of such areas does not facilitate things. In *banlieues*, spatial stigmatization is part of the daily lives of the inhabitants, youth in particular. Where they live becomes, in a negative way, a defining feature of their place in the society. As Wacquant argues, '[T]he powerful stigma attached to residence in the bounded and segregated spaces, the "neighbourhoods of exile", [. . .] is arguably the single most protrusive feature of the lived experience of those assigned to, or entrapped in, such areas' (1993: 369; emphasis removed). There is, therefore, a strong spatial dimension to the injustices experienced by the inhabitants of 'framed' *banlieues*. As many scholars working on these areas have observed, while socio-economic conditions constitute an important factor, there is also a deep feeling of injustice that leads to the explosion of revolts (Begag, 1990; Dubet and Lapeyronnie, 1992; Esterle-Hedibel, 2002; Jazouli, 1992; Lapeyronnie, 1995; Wieviorka et al., 1999). Revolts are, in this sense, unarticulated justice movements against spatial injustices (Dikeç, 2001), addressing at once material, categorical and political conditions that are spatially produced. The element of spatial injustice follows not only from economic difficulties that keep inhabitants 'trapped in space' (Harvey, 1989) or 'chained to a place' (Bourdieu, 1999), but also from the discursive articulations of *banlieues*. The remarks of Abdel from Vaulx-en-Velin are telling in this respect:

> As inhabitant, I didn't particularly choose to come and live here . . . If I could go and live elsewhere, I would. We didn't do it on purpose, all the Arabs didn't decide to come and live in Vaulx-en-Velin at Mas-du-Taureau! No more did the Africans, it's not our own doing, it's not a choice! What's so great about living here? We're not in Cannes, not in Monaco, we're not . . . we're in a *banlieue*!
> *But maybe it's the networks, you see, you know someone . . .*
> No, no, no!
> *That's not it?*
> No, that was in our parents' time. In our parents' time, when my father came, he came because he knew someone here, who'd started working, and told him, come and work, so he came, got a job, then, yes [. . .] But we didn't choose to live here . . . we chose because financial constraints make you come here, it's . . . it's other choices that make you come here.
> *Would you leave if you had the means?*
> I'd leave, if I had the means, I'd leave!
> *Do you think that's the case for most inhabitants of the neighbourhood?*
> Yeah, for lots of people, yeah . . . except for the oldest.
> *But the youngest?*
> The youngest would leave.
> *They can't leave?*
> Well, they don't have the financial means to leave, going elsewhere, when you come from Vaulx-en-Velin and when you're an immigrant, it's not possible. (Interview, Abdel Della)

As we have seen in the previous chapter, this strong spatial stigmatization negatively affects relations with employers and police, as well as with those from 'better areas' (see Wacquant, 1993). Both Abdel and Pierre-Didier, for example, relate that putting a 'better address' when applying for a job is a common – because necessary – practice. Bernard Brun, the project director in Vaulx-en-Velin, provides another example:

> We had a DSQ neighbourhood [. . .], it was called the Etats-Unis, boulevard des Etats-Unis, just that name . . . The address was boulevard des Etats-Unis, and when people wrote a cheque with that address, they were refused. So, we tried symbolic change, we changed the addresses, we created new street names, so that people, for a while at least, would not realize the person came from the boulevard des Etats-Unis, they'd live avenue, I don't know, we had made up names. (Interview, Bernard Brun)[6]

That the revolts of the 1990s occurred in the designated social housing neighbourhoods in *banlieues* does not come as a surprise in the light of these observations. What about the revolts of 2005, then? They basically shared the same geographical features, but dramatically expanded the geographies of revolts, touching 274 communes. One geographical difference was that some of the *banlieues* that were the principal sites of revolts in the 1980s and 1990s either experienced revolts belatedly in autumn 2005 (such as the *banlieues* of eastern Lyon) or stayed relatively 'calm' during the incidents (notably the notorious northern neighbourhoods of Marseille) (Lagrange, 2006a). Other than this, however, they followed a very similar geographical pattern: they occurred again in the social housing neighbourhoods of *banlieues*, most of which were the designated spaces of intervention under urban policy – the so-called ZUSs, according to the current label. Only 15% of the neighbourhoods where revolts occurred were not classified as ZUSs. The remaining 85% were urban policy neighbourhoods, and among them, the ZFUs (tax concession areas, designated among the ZUSs) were in the majority (Lagrange, 2006b). We should remember that the ZFUs are prioritized among the priority areas, seen to be having more problems, and thus have enjoyed more measures since 1996, of which tax concessions for attracting business is one. In other words, they are the sites of urban policy where public action is most concentrated.

The revolts of autumn 2005 touched two-thirds of the communes with designated urban policy areas (ZUSs). This ratio was even higher in the case of communes which had signed conventions with ANRU: 85% of the communes with designated demolition–reconstruction sites experienced revolts (Lagrange and Oberti, 2006).[7] As we have seen in Chapter 5, this programme was initiated in 2003 by Borloo with a very specific target: social housing neighbourhoods in *banlieues*, where the autumn 2005 revolts took place.

So what does this overlapping of geographies of urban policy neighbourhoods and revolts imply? Obviously, the neatly delimited areas – some of them included in urban policy for more than two decades now – do not 'contain' both the problem and the solution. Lagrange (2006b) argues, with relation to demolition–reconstruction sites, that these programmes might be creating further tensions in social housing neighbourhoods. Demolition–reconstruction means first the expulsion of inhabitants, and there is anecdotal evidence that this process is not always taking place with the involvement of inhabitants concerned, thus aggravating tensions (see Lagrange, 2006b: 112–13; and *Libération*, 24 February 2005: 7). Concerning the ZFUs, Lagrange argues that such designations may be creating expectations without fulfilling them. The utility of this measure for the inhabitants of the designated areas in terms of creating jobs is highly contested, as we have seen in Chapter 5. Thus for Lagrange, the revolts of 2005 raised 'claims aimed primarily at the state' (2006b: 122), given that they mainly occurred in the neighbourhoods of urban policy, which have been objects of public policies for more than two decades now.

While I certainly agree with these remarks, I believe there is more to it than that. First, of course, there is the spatial stigmatization that follows from such designations. When it started, urban policy did not intervene in already given spaces, but constituted its spaces of intervention as part of the policy process, consolidating, over the years, a geography of priority neighbourhoods – what I have referred to as 'the police'. This police order, as I have tried to show in Part II, has been subject to increasingly negative discursive articulations, moving from working-class neighbourhoods to 'ghettos' allegedly threatening the integrity of 'the republic', becoming almost literally a police order with constantly increasing repression.

Second, although urban policy started with the best of intentions, the heavy bureaucratic and technocratic structure it put in place, instead of encouraging local political dynamics and the right to the city – which were its initial objectives – contributed to the exclusion of inhabitants from processes that directly concerned their lives. As the *Agora* story in the previous chapter suggests, this has created further tensions by frustrating the democratic aspirations of *banlieue* youth eager to be part of the political life of their *cités*. In this sense, urban policy has, perhaps, been too present:[8]

Whereas, in the policies of which we are talking, the inhabitants are absent. What exists is rather policies of assistantship. So, it's a policy of assistantship, with respect to animation, to local policies, to projects in the neighbourhoods, that is, we have . . . from my point of view, it's a tendency to de-responsibilize people. Rather than make people act by themselves, and so on, it tends to kill initiative. Besides, when you read Dubedout, [. . .] he wrote this book, *Making the City Together*, something like that. [. . .] Well, he says clearly: you must be careful! This is what he said in the early eighties. He said you had to be careful because a neighbourhood with a policy too present, at best will result in revolts, permanent ones, and in the worst case in indifference. And that's the case. (Interview, Pierre-Didier Tchétché-Apéa, Vaulx-en-Velin)

Finally, we should bear in mind that such designations also came with increased repression – more police forces, constant surveillance, Local Security Contracts, *Maisons de Justice et de Droit*,[9] flash-ball guns, and so on. Police repression is somewhat 'targeted' on certain areas classified as 'sensitive neighbourhoods' (see *Le Monde*, 20 April 2002: 11).[10] There is, then, another layer to the overlapping geographies of unemployment, stigmatization, urban policy and revolts: geographies of repression.

Geographies of Repression: 'Police Everywhere, Justice Nowhere'[11]

We have seen that the revolts of autumn 2005 had two common features with those that occurred in the 1990s: they took place in social housing neighbourhoods, practically all of them in *banlieues*, and almost all of these neighbourhoods were among the priority neighbourhoods of urban policy. Let us go back again to the revolts of the 1990s to highlight another common feature they share with the revolts of autumn 2005, starting with the observations of Lucienne Bui-Trong, the creator of the 'Cities and *Banlieues*' section at the French Intelligence Service (RG):

And what do you think are the major reasons for the riots, why do they occur?
Riots, according to my observation, riots occur in neighbourhoods with a
    large population of immigrant origin, so they primarily reflect a difficulty
    of integration, and resentment, so, a resentment very strongly felt by young
    people of the second generation, and even the third generation too. [. . .]
    These problems are experienced as a rejection from society, and, let's say,
    they have the feeling they're relegated. But I think the riots, the context,
    the general background in which they appear, is this background, this
    feeling, this impression the people have of being relegated from society.
    [. . .] That's why in the neighbourhoods, which are also targeted by urban
    policy, in neighbourhoods that are very poor, but in which foreign popula-
    tion is not important, we don't have these phenomena of riots, because
    you don't have the same resentment against society in general, so that
    factor of riots is connected to, is connected to the fact that, one is in touch
    with other cultures, while also integrated in French culture, but with the
    feeling of being rejected by French society, you see? That's it. Now, inci-
    dents that trigger riots, that's another issue completely, you see, there's
    the triggering incident, and there's the background that's going to make
    it, because incidents triggering riots are like the spark that sets fire to

a stock of gunpowder, but that's what the gunpowder is. It's that resent-
ment. (Interview, Lucienne Bui-Trong; see also Bui-Trong, 2003).

The triggering incidents – spark to the gunpowder – constitute the third
common feature of the revolts of the 1990s and 2005. The majority of the
large-scale revolts of the 1990s (34 out of 48) were provoked by the killing,
accidentally or not, of a young person (second- or third-generation immigrant) of the neighbourhood in question. In more than half of the triggering
incidents of revolts (29 out of 48), the police were implicated (questioning,
wounding or killing). This number, however, could be higher than is suggested by the list provided by Bui-Trong. For example, the triggering
incidents for the 1991 revolts in the Val-Fourré neighbourhood of Mantes-
la-Jolie (a *banlieue* to the north-west of Paris) is given as a dispute over
entrance to a reception given at the municipality's ice rink. There is,
however, more to the story.[12]
    Following the dispute, some young people, frustrated by being rejected
entry, started attacking cars parked in the parking lot of the ice rink. The
municipality called the riot police. Confrontations with the riot police
started, store windows were smashed down, and the commercial centre was
ransacked. The riot police arrested six young people, and put them under
police custody. It was a Saturday night. One of these young people, an
18- year-old inhabitant of the neighbourhood from a North African family,
was asthmatic. He needed to regularly take medicines to prevent an attack,
and the cell where he was kept was far from ideal. Furthermore, he had
been beaten by three police officers during his arrest. Since the following
day was a Sunday, police custody was automatically prolonged as the courts
would be closed. On Monday morning, the family of the young person
brought in the necessary medicines to the police station, but they were not
allowed to give them to him since the medicines were not accompanied by
the appropriate medical certificates and necessary official authorizations.
Shortly after, the young person had an asthma attack. He was transferred
to the hospital, but too late. Spark to the gunpowder: revolts started, and
continued for two days.
    This account suggests that the police might be implicated more than the
list suggests. In addition to this, there are some curious incidents that call
into question the practices of the police. Two of the revolts, for example,
started following the killing of two people in the police station – one of
them handcuffed (which was the starting point for Kassovitz to write and
direct *La Haine*; see Favier and Kassovitz, 1995). Another started after a
police officer shot and killed a young person of African origin 'while trying
to prevent him from committing suicide'. To give yet another example: on
one occasion, the spark to the gunpowder was discharged when a 'mother
of a drug dealer [threw] herself out of the window' while the police were

searching the premises. This is not to imply that Bui-Trong's list provides false or distorted accounts. However, they evoke curiosity as to the practices of the police, and there is ample evidence of police discrimination and violence, notably towards *banlieue* youth.

The most recent example is Clichy-sous-Bois, where, as noted above, the 2005 revolts started following the killing of two young inhabitants of the neighbourhood (one of North African, the other of Malian origin), and the wounding of a third one (of Turkish origin) after they took refuge in an electricity substation to escape the police. According to the official version maintained by Sarkozy, they were not being pursued by the police, although the surviving young man, once out of the hospital, gave an account to the contrary. According to his version of events, they started running because one of the killed young men said the police was chasing them, although they were just coming back from a game of football and had done nothing wrong. They heard the sirens of the police approaching, panicked and hid in the electricity substation, where they stayed for about 30 minutes. 'I wanted to come out, go home, after all, we hadn't done anything!' But they were intimidated by the voices and barkings of dogs outside, and then it happened (see Muhittin's account in *Libération*, 16 December 2005). Let us assume that Muhittin is wrong and Sarkozy is right: even though they were not *actually* chased by the police but thought they were, what is it about the police that made these men, who had not done anything wrong according to Muhittin's account, panic and hide in an electricity substation whereas they were in their neighbourhood already? With a matter-of-fact attitude, Sarkozy never pondered the question, and stated that 'the police was not physically pursuing them' (*Libération*, 16 December 2005). Muhittin, on the other hand, was put in police custody for throwing stones at a police car during a new series of incidents in Clichy-sous-Bois at the end of May the following year (*Le Monde*, 31 May 2006).

A year before Clichy-sous-Bois, it was the *banlieues* of Strasbourg that revolted, following the 'accidental' killing of a person of North African origin by the police with a bullet in the head during a routine police road check (see, for example, *Libération*, 22 March 2004) – a form of casualty not uncommon as the triggering incident of unrest in the *banlieues*. In a book entitled *La police et la peine de mort* (The police and capital punishment) Rajsfus (2002) documents 196 deaths between 1977 and 2001, the majority of which involve youth of African or North African origin in the *banlieues*. Furthermore, the authors of such killings are usually acquitted or given very light sentences, which aggravates hostility among the *banlieue* youth towards the police, who are seen to be immune. Police violence and impunity has long been observed (see, for example, Cyran, 2003; Rajsfus, 2002), and was recently criticised openly by Amnesty International in a report on France, entitled 'The search for justice: the effective impunity of law enforcement officers in cases of shootings, deaths in custody or torture and ill-treatment'. The report, among other issues, highlighted racist police attitudes and the same geographies of repression:

> The lack of public confidence in even-handed policing is seen particularly in the 'sensitive areas' (*'quartiers sensibles'*) from which many of the victims of police ill-treatment and excessive use of force originate. Such tensions between the police and these communities have also been exacerbated when cases brought by alleged victims of police violence, or their families, eventually came to court, and resulted in highly controversial acquittals of, or token sentences, for police officers. The courtrooms, on these occasions, have been packed with friends and relatives on one side, and with police officers on the other, and scenes of violence within the court precinct have not been unknown, reinforcing the sense of 'us against them' on both sides. (Amnesty International, 2005: 1–2)[13]

A similar point was made by Patrick Bruneteaux, a researcher at the Centre for Political Research at the Sorbonne, who was interviewed by *Le Monde* on the rise of police violence in recent years: 'Dirty laundry is best washed at home. Policemen lose points or get demoted, but citizens barely ever see them being really condemned by the judicial system. It has an immediate impact in the *cités*, where violence flares up. You can hardly imagine you are in a lawful state when wrongdoers are not condemned' (*Le Monde*, 27 January 2004a).

Another issue that came to attention with the 2005 revolts was police provocation. An account provided by an eyewitness – a teacher from Clichy-sous-Bois – accused the riot police of provoking the youth by 'calling out racist insults, challenging them to fight, posturing' (Germa, 2005). The tensions arising from such provocations were aggravated when a tear-gas grenade, of the type used by the riot police, ended up in a mosque a few days after the start of revolts, and the incident was seen as a deliberate provocation by the police.[14] Another incident that took place a few days later suggested deliberate provocation by the police even more clearly. This time the place was Lyon, in the social housing neighbourhood of La Duchère,[15] and the scene was recorded by a journalist using a hidden camera. The police were conducting identity checks on a group of young men, one of whom protested. The following dialogue ensued:

Policeman: Shut up!
Young man: You tell us to shut up but we haven't done anything wrong, Sir.
Policeman: Do you want me to take you to a power transformer?
Young man: Sorry, Sir, you're not talking nice to me, I didn't talk to you, Sir.

Policeman: So don't talk to us! ... We're telling you to move back, move back!

Young man: Listen, Sir, we're addressing you respectfully [*on vous vouvoie*] and your colleague's not answering the same way [*il nous tutoie*]! We're being respectful!

Another young man to one of the officers: Well done! You have cancer! You've lost all your hair!

The officer responds: Hey, you wanna fry with your pals? You wanna go into a power transformer? You just keep going, and we'll take you.

The first young man: If you behave like this, do you really think the neighbour-hood is going to calm down?

Policeman: We don't give a shit whether the neighbourhood calms down or not. In a way, the worse the shit, the happier we are![16]

These examples show once again the tensions between the police and the *banlieue* youth, and suggest the possibility of a deliberate police provocation (see also Le Goaziou and Mucchielli, 2006). Indeed, this is more than just a possibility according to a recent report by the French Intelligence Service (RG), which points to the responsibility of the CRS (riot police) in the aggravation of tensions and ensuing revolts in Clichy and Montfermeil. As an officer from the RG explains:

We don't say explicitly that our colleagues created the problems. We avoid criticizing each other. But we have reported to our superiors that there are useless provocations. Unless what is intended is to set fire, again, to the *cités*. In this respect, mobile police squadrons are clever. They intervene more subtly. The CRS, on the other hand, are obedient. They hit first and think afterwards. In the field, I meet more and more parents, families who report violence from the police. Recently, a father and his teenaged daughter told me their house was shot at with flash-balls. With no reason. (*Le Canard enchaîné*, 7 June 2006: 4)

These are worrying developments. In 2003, police violence was on the rise for the sixth consecutive year (*Le Monde*, 27 January 2004b) – since, in other words, the turn of the left to the 'right to security' in 1997. The coming to power of a right-wing government in 2002 with a deliberate repressive policy only exacerbated the situation.

Policies of Urgency: '20 Years for Unemployment, 20 Minutes for Insecurity'

It's difficult to have a logical, general, systematic discourse about rethinking the city, while people are saying: 'Well, that's discourse and that's long term,

it's to be left to academics with leisure.' We have to answer the population's claims: 'we want security, we want police, we want cameras. It works well elsewhere, so we're following suit'. And that can be done in fifteen seconds.
Luc Gwiazdzinski, urbanist (interview in *Marchands de sécurité*, 2002)

Increasing police violence and repression, targeted more openly towards the *banlieue* youth, has to be seen in a context of an ever-increasing obsession with security in recent years. As we have seen in Chapter 5, the Socialist government's prioritization of the issue of security in 1997 and the presidential elections of 2002 were important turning points. The elections of 2002 were so marked by the issue of insecurity that an extreme right militant made the following comment: 'Now everybody talks about nothing but insecurity, immigration, and the authoritarian functions of the state. When we used to talk about these issues, we were being treated as fascists' (*Libération*, 28 December 2001: 11).[17]

It then comes as no surprise that in 2000, the fastest growing sector of the French economy, hiring and advancing more than any other, was the private security sector – the 'security merchants' ('*marchands de sécurité*'), as a documentary called them (*Marchands de sécurité*, 2002).[18] The prioritization of security in 1997 was followed by its privatization, partly at least, with the introduction of Local Security Contracts (CLSs). Since then, 'the *banlieue* is the new El Dorado' for private companies offering security services, such as the preparation of 'local security diagnostics' (a prerequisite for having a CLS), creation of municipal police and installation of surveillance cameras (*Marchands de sécurité*, 2002). As the 'urbanist' whose remarks open this section argued, such measures would have been 'unthinkable' five or ten years ago:

> I was lucky enough, a few years ago, to go to the US and the UK to work on these issues of security policy, and I came back with a few experiences in mind. I came back with the experience of curfews: 280 cities in the US had set up curfews for teenagers. I came back with the experiences around cameras, camera systems to control inner cities in particular. I also came back with those ideas and policies set up around the private cities, the gated communities. I also came back with these expanding ideas about private police, the privatization of private [*sic*] police, the development of private security, with the development of Giuliani's zero tolerance policy in New York, too. All these things, ten, five years ago, would have been unthinkable in France. Now they exist. (Luc Gwiazdzinski, interview *in Marchands de sécurité*, 2002)

Zero tolerance has been an explicitly stated policy of the government since 2002, the effects of which were particularly felt in the social housing neighbourhoods of banlieues. As Lahlou of the MIB puts it:

The only response to people's misery, nowadays, is repression and imprison-
ment. They think the prison is the answer. [. . .] There's no zero tolerance
policy against people being hit by trouble, say. I mean, there's this social insecurity, no one talks about it: unemployment, lousy housing, people deep in
shit, excluded from the social arena, with just a minimal income, or even
less . . . No one talks about that, because, since they have no solutions, they
prefer to reassure public opinion about zero tolerance [for crime] so once again,
who gets hit? The *banlieue* gets hit, necessarily. (Interview, Lahlou, MIB)

We have already seen in Chapter 5 how the government rapidly introduced
many repressive measures immediately after taking office in May 2002 – so
rapidly that a young inhabitant of Vaulx-en-Velin commented as follows:
'They couldn't find the solution to unemployment for 20 years now, but
they have found the solution to insecurity in 20 minutes.' This hard-line
policy only intensified in the following years, again with a spatial focus on
the social housing neighbourhoods of *banlieues*. In January 2004, Sarkozy
held a press conference to define his priorities for the year, and announced
'a merciless fight against urban violence and parallel economies' (which
means drugs in the *banlieues*). One priority involved a reform of the French
Intelligence Service in order to increase its efficiency in fighting against
terrorism and 'urban violence'. Another was even more explicitly spatial.
Sarkozy stated that the 'rule of law' ('*état de droit*') would be restored in 20
spatially targeted 'outlaw areas' ('*zones de non-droit*') before the end of the
year (*Le Monde*, 14 January 2004; *Libération*, 15 January 2004; see also *Le
Monde*, 16 January 2004 for the reactions of the inhabitants of Val-Fourré,
which was on Sarkozy's list). The definitive list of communes was announced
a few days later, and included 23 communes with 'sensitive neighbourhoods' (*L'Humanité*, 27 January 2004). All of these were included in urban
policy (four since 1982, nine since 1984, eight since 1989 and two since
1996), and more than half experienced revolts in autumn 2005.

Sarkozy's zero tolerance policies were premised on what he called a
'culture of results' ('*culture du résultat*'), which meant 'better' numbers (i.e.
more police actions and less delinquency). This 'culture' imposed on the
police by Sarkozy was seen as partly responsible for the increase in police
violence by the organizations defending human rights (which were, by the
way, ridiculed by Sarkozy with the expression 'human rightists' ('*droit de
l'hommistes*') (*Libération*, 6 May 2004)). Sarkozy wanted immediate results
so much that when a group of proximity police officers in Toulouse told
him they had organized a football game with the youth of the 'sensitive
neighbourhood' of Bellefontaine (which was on Sarkozy's 'black-list'), he
scolded them before TV cameras: 'You are not social workers. The best of
preventions is sanctioning' (*Le Canard enchaîné*, 5 March 2003: 4).

This emphasis on 'results' also produced curious practices among the police. In April 2003, for example, a police chief sent a letter to his 140 police officers asking them to 'boost the figures', for the statistics in his district were not good enough. He wrote: 'The technical control of the district currently in progress shows an obvious *deficit* in terms of elucidation rates on which the effort has to be increased in order to allow the presentation of statistics in conformity with the departmental average.'[19] The stress grew as the end of the year approached, since the Minister of the Interior publicized the figures at the beginning of each year, and those bringing in 'bad' figures risked their jobs. Thus, in October 2003, an even more curious letter was circulated by the public security director of the department of Hérault (in the Languedoc-Roussillon region). Reckoning that the numbers in his department were not satisfactory, the director sent the following instructions to his officers: the anti-criminal brigade '*must* [emphasis in original] achieve the minimum objective of six police custodies per day'; the daily shift a 'minimum four police custodies per day' of which a 'minimum two at night'; only two for the dog brigade, and seven for the proximity police, not forgetting at least one by the road team. Funny though it may seem, the letter, in fact, constituted a violation of the penal procedure, which the secretary of the national police officers' union interpreted as resulting from the emphasis placed on results by the Minister of the Interior (*Libération*, 31 October 2003).[20]

These developments and such practices are important in order to make sense of the revolts of 2005 in the *banlieues*. Another important point to keep in mind is the Minister of the Interior's use of inflammatory language towards the *banlieue* youth, which definitely did not help to calm things down. Three months before the revolts, on a visit to an emblematic *banlieue*, the *cité des* 4000 in La Courneuve, Sarkozy had talked about 'cleaning the *cité* with Kärcher' – a well-known brand of power hoses for cleaning surfaces through sand- or water-blasting.[21] During the revolts, he referred to the revolting youth as '*racaille*' – a rather pejorative term usually translated as 'scum' or 'rubble' – and proposed the expulsion of foreigners (including those with residency permits) implicated in the incidents.[22] The insults did not end there. On 10 November, while the revolts still continued, Sarkozy was invited on a TV programme on France 2 about the *banlieues*: 'They are thugs [*voyous*] and scum [*racaille*], I'll stick to my guns.' Once the revolts were over, he would regret using the term '*racaille*', but not because it was overly pejorative – to the contrary, it was too 'weak' a term to qualify the revolting youth:

And honestly, if I regret one thing, it's to have used the term *racaille*, which is way too lenient if you look at the judicial pedigree of some of the individuals arrested during the riots. It's the law of the republic and not the law of the gangs that prevailed. (*Libération*, 21 November 2005: 13)

The use of this kind of inflammatory language is not new – Chevènement, for example, used to talk about 'little savages' ('*sauvageons*') – but Sarkozy definitely raised the bar, adding to the stigmatization of the *banlieue* youth. I have tried to show some of the problems that the inhabitants of *banlieues*, youth in particular, have to face on a daily basis – unemployment, discrimination, stigmatization, police violence and an increasingly hard-line, even insulting, official discourse against them. But this is not the image of the *banlieues* constituted by the state's statements of the recent years, not to mention the media. Sarkozy's assertion is exemplary here:

The primary cause of unemployment, of despair, of violence in the *banlieues*, it isn't discrimination, it isn't failure of the educational system. The primary cause of despair in the neighbourhoods, it's drug trafficking, the rule of gangs, the dictatorship of fear and the abandonment by the republic. (*Le Monde*, 22 November 2005: 12)

This brings to mind the remarks of an activist:

They say, 'but these are not social issues, this is about public order'. So what happened was that from the phrase of the 1980s, 'neighbourhoods in danger', to be taken care of, they shifted to 'dangerous neighbourhoods', and there you go. (Interview, Guy, MIB)

Conclusions: From 'a Just Revolt of the Youth' to 'Urban Violence'

Despite similarities with the previous large-scale revolts in the social housing neighbourhoods of *banlieues*, the revolts of 2005 were nevertheless unprecedented in terms of their magnitude and persistence. The measures used to repress them were unprecedented as well, including the declaration of a state of emergency, allowing curfews to be imposed, on 8 November 2005 – just when calm was returning to the *banlieues*.

The government's response to the revolts was marked by a characteristic concern with rapid and increased repression, and in this sense, it differed remarkably from the responses of previous governments. When faced with the 1981 revolts in the *banlieues*, the Socialist government of the time took the incidents seriously and initiated an urban policy programme with strong political ideals. Following the 1990 revolts, a City Ministry was created as a sign of the commitment of the state to the 'urban question'. But the 1990s also saw the consolidation of a particular spatial order constituted mainly by the social housing neighbourhoods of *banlieues*, and gave the first signs of the coming of the penal state with the involvement of the Ministry of Justice and the French Intelligence Service (RG). The involvement of the RG, in particular, was a very strong statement. The secrecy of the activities of the RG, and its more traditional focus on terrorism strengthened the impression that the *banlieues* were 'threats' to the French society. Thus, not only was the constant surveillance of the *banlieues* justified, even rendered necessary, but new ways of talking about them were generated with the appearance of new notions such as 'urban violence', 'sensitive neighbourhoods' and 'urban guerrillas'.

The RG's involvement was also effective in the production of new statistics on the first of these terms, 'urban violence', which entered the agenda in the early 1990s.[23] Since no statistical information was gathered before to measure 'urban violence' – the category simply did not exist – the impression was that there was a sheer explosion of urban violence in France. Thereafter, these statistics were regularly used by the media and politicians asking for more security measures.[24] However, Lucienne Bui-Trong of the RG was very clear on the production of these statistics. Starting from 1991, the RG had asked the departments to provide them with 'information on difficult neighbourhoods', which would then form the database of the section on 'urban violence' (the early name of the section 'Cities and *Banlieues*'). As the statistical category did not exist before, there was indeed a sheer explosion of 'urban violence' in France in the early 1990s. And 'urban violence' kept increasing not necessarily because there were more acts of 'urban violence' (however defined) every year, but because more departments were concerned, more neighbourhoods were included, and more incidents were *reported*. The number of surveyed neighbourhoods increased threefold in five years, thus increasing the quantity of reported incidents as well. Furthermore, some departments were more eager and capable (that is, with more agents working on the issue) than others to provide information on 'urban violence' for the RG's database, as the following remarks make clear:

> In 1991, we placed an order with the departmental directions to spot the difficult neighbourhoods [*quartiers difficiles*]. And the civil servants managed to get us the information. Originally, those agents were not specialized, obviously; they worked traditionally on political or social issues, as generalists on a geographical area, or they were specialized on a single issue. At the RG, missions are very diverse depending on departments, because we are a small service: according to local situations, each departmental director manages with the means at his disposal. The first studies were done by generalists who had never worked on this issue in particular but who had been given a guideline for their research. Little by little, people in the services took on this new mission. Sometimes specialists were designated. In some departments, there

are four or five people working exclusively on the difficult neighbourhoods; in others, a single civil servant spends a few hours a month on them. (Bui-Trong, 1998b: 227–8)

The status of delinquency statistics, which is also highly popular in the media and among politicians, is no less problematic. There is, indeed, a major flaw in the production of the statistics both of delinquency and of 'urban violence': they are the results of the activities of the police. Furthermore, unlike other institutions, say, the INSEE or INED, charged with collecting statistical information on issues that have nothing to do with their own activities, statistics of delinquency are collected by institutions that are directly charged with the issue – the Ministry of the Interior and the Ministry of Justice. Thus, activities of the police become a mirror-image of delinquency. And here lies the flaw: 'If the repressive priorities of the police change, its forces increase, if the minister issues strict orders, the statistics at the end of the year will show evolutions that bear no relation with the evolution of criminality. [. . .] However, every year the figures and the interpretations of the ministry of the Interior are directly quoted by the media' (Mucchielli, 2001: 24–5).

This production of new terms and the putting in place of sensible evidences such as statistics – however flawed – re-configured a perceptive field around the *banlieues*, which increasingly associated them with violence and delinquency. This re-configuration, which started in the early 1990s, has largely contributed to the debilitation of the political significance of revolts. In the 1980s, revolts were seen as having a political significance, an interpretation shared by Lucienne Bui-Trong as well:

The really critical events were the violences which took place in Vénissieux, at the Minguettes, during the summer, the summer 1981 I think, but it was, they were, those violences, phenomena of degree 5 or 6, on my scale. But still, it had already considerably impressed the ministry, well, the government, *as it was a left-wing government*, so that government was looking for very comprehensive solutions, and from there they launched the, the Primer Minister Pierre Mauroy had asked Bonnemaison, the commission of mayors, to prepare a whole doctrine on that, so, it was the starting point for the police, for urban policy, let's say, a social and comprehensive work on the neighbourhoods. (Interview, Lucienne Bui-Trong)

This vision, however, started to change starting from the 1990s, as we have seen in Chapter 4, and the line that separated the left from the right in terms of repression eventually disappeared towards the end of the decade, as we have seen in Chapter 5:

In 1991, when the government wanted to talk only of a 'prevention police', and interpreted violence [i.e. incidents of 1990 and 1991 in *banlieues*] as '*a just revolt of the youth*', I emphasized the unease of the police faced with this discourse; I tried to show that violence in itself was a problem. [. . .] The first task of my section, therefore, had been to make known in higher places the policemen's point of view on the question of *banlieues*, their discontent with the way the phenomenon was being treated ideologically and in the media. (Bui-Trong, 1998b: 230 and 227; emphasis added)

This change from one decade to the other has also been observed by sociologist Eric Macé:

But the difference between the 1990s and the 1980s is that, in the 1980s [. . .] there was a political awareness, saying: 'Deep down, these violences have a political significance. They challenge French society on its ability to integrate generally.' In the 1990s, under the term 'urban violence', what constitutes a threat is designated – what threatens, and what has to be intervened in, in fact, to protect us against this threat. (Interview in *Marchands de sécurité*, 2002)

Such a change in the perception of the *banlieue* revolts has largely shadowed the political significance of such incidents. With the re-configuration of a perceptive field around *banlieues* with such terms as 'urban violence' and 'outlaw areas', it is not surprising that repression has become a focus – and 'legitimately' so. This re-configuration highlighted less the difficult material conditions in the *banlieues* than the 'threat' posed by 'the *banlieue*' to security, social order and the integrity of 'the republic', rendering episodic manifestations of discontent acts of violence rather than claims for justice – turning, in other words, voices into noises. By confining the 'other' within a geographical elsewhere, by closing the *banlieue* in on itself, this consolidation of the police order not only removed from perspective the structural dynamics of persistent inequalities and injustices, but also re-configured the 'givens' of the situation by constituting 'the *banlieue*' as a problem in itself, treating the claims rising from the *banlieues* not as voices that question the order of things, but as noises that disturb the established order. However, overlapping geographies of inequalities, discrimination, police violence, repression *and* revolts suggest another interpretation. The geographical pattern and expansion of revolts imply that there are structural dynamics aggravating inequalities, which particularly hit the social housing neighbourhoods in *banlieues*. Revolts, therefore, are not just looting and burning; even though they are marked by elements of violence, they connect and speak to larger dynamics and severe material conditions. They are, in this sense, unarticulated justice movements.

8

Conclusion: Space, Politics and Urban Policy

I began with a central premise to consider space not as given, but as produced through practices of articulation, such as policy discourses, namings, mappings and statistics. My attempt, however, has not been to cover the various practices that produce spaces, but, rather, to consider some of the implications of my central premise for looking at a particular articulatory practice – urban policy. While the book's scope necessarily extended beyond urban policy, the spaces urban policy constituted and intervened in remained central to its analysis.

Yet those spaces themselves have gone through remarkable changes over the years. As we have seen in chapters 3 through 5, they have been subject to different articulations, which eventually turned them into 'threats'. Despite the fact that urban policy intervened in the same physical spaces for years, it articulated them in different ways, changing the discursive terms with which to talk about them, the formulation of problems, the proposed solutions and the legitimation of state intervention.

In my attempt to develop an approach to urban policy with a focus on space, I tried to interpret these changes in the light of my central premise to consider space as produced through articulatory practices. Urban policy, I have argued, does not intervene in given spaces, but constitutes them as part of the policy process. This implies giving due attention to various practices of articulation involved in the constitution of spaces as objects of urban policy interventions, such as policy documents, mappings, categorizations, definitions, spatial designations, descriptive names and statistics – what I have referred to as 'the state's statements', following Corrigan and Sayer's (1985) formulation. In this view, urban policy – as a statement – is not simply a technical or administrative tool to deal with perceived problems in space (or 'problem spaces'), but an institutionalized practice of articulation that actively constitutes space. This is what I have referred to, in Chapter 2, as urban policy as place-making. Spatially designating areas, associating problems with them, and generating discursive terms to talk about them, urban policy, I have argued, is a place-making practice which affects the daily lives of inhabitants concerned, as we have seen in Chapter 6. But I have also argued that this practice does not go uncontested, an issue I will address shortly.

As we have seen, changes concerned not only the discursive articulations of the spaces of urban policy, but their conceptualizations as well. By highlighting such changes, I have tried to show that space matters in urban policy. Saying this sounds a bit like stating the obvious; after all, urban policy is about space, or, better yet, about certain spaces. It defines the issues it seeks to address in spatial terms, delineates spaces to intervene in, and proposes spatially targeted remedies to perceived problems. Yet, as we have seen in Part II, spatial conceptualizations change. Urban policy does not operate with an unchanging view of space. Just as the discursive terms used to designate its spaces change, the way it conceptualizes its spaces changes as well. In other words, urban policy is guided by particular ways of imagining space, and different ways of imagining space have different implications for the constitution of perceived problems and proposed solutions. Depending on whether spaces of intervention are conceived as neatly delimited areas or in relational terms, initiatives range from limited local interventions to regional distributive policies.

Changing spatial conceptualizations have another implication as well. In Chapter 2, I have related urban policy to the notion of 'state's statements', which allowed me to highlight the governmental dimension of urban policy. However, while suggesting that the place-making practices of urban policy have a governmental dimension, I have insisted that this is far from univocal. Governmental practices are not about interventions only; they involve both formation *and* intervention. If we hold the view that space is not already given in urban policy, then it follows that urban policy is a governmental practice not merely because it intervenes in certain spaces, but because it constitutes them as objects of intervention in the first place. In this sense, the spaces of urban policy are not the given sites of this governmental practice, but, first of all, its outcomes.

But as we have seen in Part II, the ways in which such spaces are imagined and used in urban policy programmes vary, which was perhaps best exemplified by the sharp differences between the early – militant – years of urban policy and its later – statist – orientations. So while it is important to recognize the governmental dimension of the constitution of spaces through urban policy, it is equally important to remember that there is no inherent politics to such constitutions.

This brings me to the issue of space and politics. I have argued that urban policy depends on particular spatial imaginaries. But it also deploys them, consolidating a spatial order that becomes the basis of its interventions. Following Rancière, I have referred to this consolidated spatial order – provocatively perhaps – as 'the police', understood in its non-pejorative sense as organization and government. The notion allowed me to highlight the governmental dimension of urban policy since the police refers to a symbolically constituted organization of social space – which may also be stubbornly inscribed in physical spaces, as the *banlieues* exemplify – which becomes the basis of and for governance. It also enabled me to signpost the shape of things to come; the police order consolidated by urban policy, I have attempted to show in Chapters 5 and 7, has become almost literally a police order, with ever more repressive measures.

Rancière's conceptualization of the police resonates well with my central premise to consider space as produced through practices of articulation. His insistence on the contingency of the police order, produced through a particular regime of representation – through, in other words, the putting in place of 'sensible evidences' – enabled me to interpret urban policy as a regime of representation that consolidates a certain spatial order through various practices of articulation (spatial designations, namings, categorizations, mappings, statistics, and so on). Arguably, the most perverse consequence of the consolidation of this spatial order – the police – has been the constitution of '*banlieues*' as spaces that somehow do not fit, excluded, dangerous, deviant – as, in other words, a form of exteriority that menaces the integrity of 'the republic'.

But nothing escapes the police, especially 'the excluded'. Politically, identifying 'the excluded' as the excluded is to already include 'them' in the police notion of the whole to be governed. Space, I have argued in Chapter 2, is relevant to the police because the police depends on the partitioning of spaces and distribution of places, names, functions, people, activities, and so on. Naming, fixing in space, defining a proper place, are all functions of the police, whose principle is not repression but distribution.

This partitioning of space by the police has two governmental implications. On the one hand, the consolidation and normalization of a certain police order provides the basis of and for governance. We have seen in Chapter 4 how the spatial order consolidated by urban policy provided the basis for the activities not only of urban policy, but also of the Ministry of Justice and the French Intelligence Service – not to mention political discourses and debates in the media, which increasingly used that spatial imaginary when playing with xenophobic and authoritarian sensibilities.

The consolidation of the police, on the other hand, arranges the perceptive givens of a situation – defining what is in or out, distinguishing 'the excluded' from 'the included', the audible from the inaudible, the visible from the invisible. In Chapter 7, I have attempted to show how the police order highlighted less the difficult conditions in *banlieues* than the 'threat' posed by them, contributing thus to the gradual disappearance of the political significance of revolts. The police order, therefore, not only becomes the basis for government, but – once normalized – it also provides a particular locus of enunciation for the governed. Remember the remarks, in Chapter 2, of Yves Mena from Vaulx-en-Velin about the difficulties generated by 'the place' of *banlieues* in the consolidated police order: 'you're a bit too connoted'.

So far I have focused on the role that urban policy (and other state's statements) plays in the constitution of spaces and the consolidation of the police. Urban policy is one way of constituting space, but it is not the only one. The process through which urban policy and the state's other statements produce spaces is not uncontested. There is, in this sense, a contestation for space, which I have tried to highlight by remarks from local activists throughout the book, and by focusing on a particular case in Chapter 6. To put it in the language of 'the police', I have argued that the spaces of the police and spaces of politics are enmeshed. What the state's statements constitute as 'badlands' are also sites of political mobilization with democratic ideals, opening up the saturated spaces of the police as spaces of politics. Resistance to the police is as much about place-making as the police itself; the confrontation of the police and politics, in this sense, is a contestation for space. The remarks of Lahlou, one of the founding members of the *Mouvement de l'Immigration et des Banlieues* (MIB), are exemplary here:

> The *banlieue* has been, and still is, so decried, you know, by the power or even the media. Being young and from the *banlieue*, inevitably it is very bad when you live in the 93 [the number of Seine-Saint-Denis, the most industrial and working-class department of the Paris agglomeration]. They say, if you apply for a job, after going to university, you shouldn't say you're from the 93, it's very pejoratively connoted. So we, on the other hand, wanted to say so explicitly, you see. Our pals said so, the girls who were there said so . . . *banlieue* was also regarded in a very negative sort of way, as I mentioned, you know. Politicians, both left- and right-wing, and some media that were very complicit in the whole thing, they carried the message, spread it around, always in a negative way. Which means you're constantly faced with a total negation of yourself, you know, your personal history and whatever . . . And then, we, the MIB, we were going to try, we're still trying, to spread a different image, to say, 'Hey, stop this whole trip, the northern *banlieue* has nothing to do with that.' (Interview, Lahlou, MIB)

These are some of the lessons offered by French urban policy when approached with a focus on its politics of space. Another lesson offered by the French experience is that urban policy cannot be understood independently of the broader political and economic context. It has to be understood in a range of established political traditions – French republicanism, in this case – and major national and international events – from riots in Brixton to student demonstrations in Paris, from the Rushdie affair to the Islamic headscarf affair, from the Intifada to the Los Angeles riots.

Furthermore, urban policy has to be seen in relation to broader state restructuring, paying due attention to dominant rationalities and legitimizing discourses as the state's statements in the domain of urban policy are not produced in a vacuum. The dominant discourse of recent years, as we have seen in Chapter 2, has been about the neoliberalization of North American and Western European states, and the role urban policy plays in this process by creating the conditions for making the production of competitive spaces a priority, even a necessity. The chapters in Part II showed, however, that although the French state is committed to some form of moderate neoliberalism in its contemporary restructurings, the effects of this commitment on urban policy have been only partial. Despite experimentations with neoliberal ideas, French urban policy has not sought to institutionalize inter-urban competition and to encourage a growth-first competitive logic as an overriding goal. Economic growth and competition have not replaced social issues as primary objectives; urban policy is a social, not an economic development, policy, although recent developments point to a diminished emphasis on social issues on the part of the state. In this sense, urban policy has followed the political rationality of the republican tradition emphasizing the social obligations of the state, and not that of neoliberalism, which seeks to extend and disseminate market values. This suggests that accounts of neoliberalism and state restructuring should be sensitive to established political traditions, which implies giving due attention to various political rationalities guiding policy-making, inherited institutional structures and practices, and the role of dominant political traditions in shaping both state actions and the public response to them.

Thus, while French urban policy has not become a means for producing competitive spaces, it has become more and more concerned with containing certain spaces and populations seen to be problematic, reflecting the contemporary transformations of the French state along more authoritarian and exclusionary lines – which I have tried to capture with the notion of 'republican penal state' in Chapter 2. An integral part of the consolidation of the republican penal state has been the state's statements that constitute spaces and consolidate police orders. As we have seen in Chapters 5 and 7, in particular, urban policy and its spaces of intervention have become the major sites upon and through which the republican penal state consolidated itself. The legitimizing discourse of this consolidation has been centred on 'the republic', allegedly under threat by the formation of 'communities' at its gates, incompatible with its values and principles. The disciplinary practices of the penal state, then, are substantially spatial practices that produce spaces of intervention (or containment), although the discursive articulations of such spaces and modes of legitimization vary depending on forms of state restructuring.

I have argued that established political traditions matter in urban policy (and in state restructuring, more broadly). I have insisted, however, that such traditions be treated carefully, paying attention to their variations and often paradoxical interpretations, rather than taking them as a univocal and stable 'model' with coherent policy implications (see Chapter 2). In this sense, the response of the government to the autumn 2005 *banlieue* revolts was exemplary, showing not only the contemporary transformations of the French state along increasingly authoritarian lines, but also the paradox of 'actually existing republicanism', with which I would like to conclude.

The government's response to the revolts of autumn 2005 was marked by a concern with rapid and increased repression, therefore not deviating from its previous policies since 2002, which we have seen in Chapter 5. Just when calm was returning to the *banlieues*, a state of emergency was declared, allowing curfews to be imposed. This was an unprecedented response to revolts in the *banlieues*, and the fact that the state of emergency was based on a 1955 law dating from the Algerian war only added insult to injury.[1] It certainly brought to mind what was so bluntly expressed by Abdel of Vaulx-en-Velin: 'the Algerian war is not over in France' (see Chapter 6).

As we have seen in Chapter 1, the term *banlieue* has always been negatively connoted, although the terms of this negative connotation have gone through changes over the years. Starting particularly with the 1990s, there has been a strong stigmatization of *banlieues* with references to the formation of ghettos, ethnic separatism (*'communautarisme'*) and Islamic fundamentalism. When the 'threat' of *banlieues* was articulated in the 1970s, it did not involve 'ethnic' and religious connotations. Starting with the 1980s, however, the *banlieues* were closely associated with the 'problem of immigration', the problem being the 'integration' of non-European immigrants and their descendants into French society. Then, starting with the 1990s, Islam became a dominant theme.

This change was not simply a straightforward reflection of the shifting demographic composition of the social housing neighbourhoods in the *banlieues*. The context in which the major *banlieue* revolts were articulated largely contributed to the consolidation of the current negative image of *banlieues*. In the early 1980s, the Socialists were in power for the first time in the Fifth Republic with a politically contentious agenda, which included, among others, the suspension of the expulsion of immigrants, and voting rights for immigrants in local elections. The revolts of the 'hot summer' of 1981 took on a specific importance in this context, where the opposition right had focused its critique on the 'soft' attitude of the new government towards immigration. The revolts of 1990 occurred in a context of heated debates around immigration and Islam, marked by such incidents as the Islamic headscarf affair of 1989 in France, the Intifada (already in place for three years), the Rushdie affair and the Gulf War (to start in January 1991). Similarly, the revolts of 2005 occurred in a particularly tense context, marked by September 11 and its aftermath of increasing preoccupations over terrorism, security and Islam. Other incidents – such as war in Afghanistan and Iraq, the bombings in Bali and on the London tube, and debates around the entrance of Turkey to the European Union – contributed to the development of hostile arguments against non-European immigrants and Muslims in many Western states, including France, where the *banlieues* became the spatially reified forms of such 'threats'.

It is important, therefore, to see the current negative image of *banlieues* as articulated in and with reference to these particular contexts rather than as an unproblematic reflection of reality. This articulation, as I argued in Chapter 7, highlights less the difficult material conditions in *banlieues* than the 'threat' posed by *banlieues*, shifting focus from growing inequalities and discriminations to menaces to 'the values of the republic', French identity and the authority of the state. Furthermore, this articulation constitutes the *banlieues* in the form of a *menacing exteriority*, which not only makes the application of ever more repressive measures possible, but also largely debilitates the significance of recurrent revolts.

The 2005 revolts, in this sense, were no exception, and they quickly gave rise to debates about 'integrating' the children of (non-European) immigrants, ethnic separatism and Islamic fundamentalism. In terms of their magnitude, they were compared to the May 1968 uprisings. Broadcast images of the revolting youth mainly showed 'darker' people, which led to the interpretation of revolts as 'ethnic' and 'religious' – an interpretation much exploited by the media in the United States (Fox News, for example, reported the revolts as 'Muslim riots'). Let us follow the May 1968 analogy: stating that the revolts were 'ethnic' (dark skin) or religious (Islam) is almost as absurd as stating that the May 1968 uprisings were 'ethnic' (white) or religious (Christian). There was nothing to suggest that the revolts were 'ethnic' or religious. No such claim was made. *Banlieues* (the 'bad' ones) are not particularly attractive places in which to live. Many of them are marked by deteriorating housing, lack of facilities, education problems, insufficient transportation and a strong territorial stigmatization, the negative effects of which are strongly felt on a daily basis by their inhabitants, youth in particular (for example, in relations with the police and job applications, as we have seen in Chapter 6). If the majority of revolting youth have darker complexions, this should raise questions about the dynamics of the housing market and the practices of social housing construction, distribution and allocation, rather than lead to a hasty conclusion that it is the Arabs, blacks and Muslims who are rioting. Rather than the 'ethnic' origins or alleged religious affiliations of those who revolt, attention should be given to the fact that revolting geographies, as I have sought to show in Chapter 7, are also geographies of inequalities, discrimination and repression.

Therefore, suggesting that the revolts of 2005 were 'ethnic' or religious is misleading in terms of the dynamics behind revolts and their political implications. Confining the *banlieue* youth (most of whom were born and/or raised in France) within already given 'ethnic' or religious identities is to already place them on the outside, to be 'integrated', it is hoped, by the so-called 'republican model of integration'. As a result, when the youth in *banlieues* revolt, they always 'revolt as' (as blacks, as Arabs, as the children of immigrants, as Muslims). This is not to deny the racialized basis of inequalities and discriminations, but to point to the perils of confining political subjectivities within already given categories of identity. Rather than confining their spontaneously constituted political identities to pre-conceived categories that are deemed incompatible with the principle of the 'one and indivisible republic', the challenge, it seems to me, is to resist the place assigned to them in the police order, and to hear their voices as equals manifesting their discontent and desire to be treated as equals.

The problem is not that republicanism is inherently incompatible with diversity. The problem is that the republican imaginary is so white and so Christian that any manifestation of discontent – either on the streets or in the spaces of institutional politics – by the republic's darker and non-Christian (or thought to be so) citizens quickly evokes concerns about the values and principles of the republic. This is the paradox of actually existing republicanism in France. When those who do not quite fit in the republican imaginary mobilize, the principle of equality – otherwise strongly defended – gets displaced by a preoccupation with 'ethnic' origins and religious affiliations – otherwise strongly criticized. Rather than a defence of the equality of all its members regardless of ethnicity or religion, republicanism becomes a denial of diversity, and prevents the constitution of political spaces where the voices of the youth in *banlieues* can be heard *as equals* manifesting their discontent, making a claim on the republic as part of the republic – not as barbarians at its gate.

Notes

CHAPTER 1   INTRODUCTION: THE FEAR OF 'THE *BANLIEUE*'

1 Inhabitants of a 'zone', a term I explain below.
2 The term dates from the medieval era, and is formed by the combination of '*ban*' (official proclamation of something, in particular an order or an interdiction) and '*lieue*' (one league, about four kilometres). It was used to designate the area surrounding a city over which the '*ban*' extended. See the entry in Merlin and Choay (2000).
3 Another example would be the expression '*être/mettre au ban de la société*', which means 'to be outlawed/to outlaw from society'.
4 This may be exemplified by an article that appeared in the daily newspaper *Le Monde* on 1 March 2002 (p. 9), which carried the following title printed in large type at the top of the page: 'The police faced with the elusive underground economy of *quartiers*'. It becomes clear, in the body of the article, that the 'underground economy' in question is organized mainly around drugs, and the term '*quartiers*' is used not to refer to any neighbourhood, but to 'neighbourhoods in difficulty' ('*quartiers en difficultés*', a term institutionalized by urban policy, which designates its selected neighbourhoods) and the *banlieues* more generally.
5 Sites of insurgence 'introduce into the city new identities and practices that disturb established histories' (Holston, 1998: 48). Such sites, in Holston's view, are the sites of 'insurgent citizenship' – various practices that range from everyday activities to grassroots mobilizations – that bring into question the meaning of membership in the modern state.

CHAPTER 2   STATE'S STATEMENTS: URBAN POLICY AS PLACE-MAKING

1 Vaulx-en-Velin is located in the Lyon metropolitan area. More on this *banlieue* and its local activists will be presented in Chapter 6.

2 As Rancière explained in an interview: 'If, among the thinkers of my generation, there was one I was quite close to at one point, it was Foucault. Something of Foucault's archaeological project – the will to think the conditions of
possibility of such and such a form of statement or such and such an object's
constitution – has stuck with me' (Rancière, 2003a: 208–9).

3 In the English translation, this part is given as 'a place's destination'.
'Function', I believe, would be a more appropriate translation, which is what
is implied in the phrase '*la destination d'un lieu*' (see Rancière, 1995: 53).

4 Although I focus mainly on the practices of the state, it should be noted
that Foucault's notion of governmentality did not refer merely to the state's
practices.

5 The CPE was aimed at young people aged under 26, and it sought to encourage the creation of jobs in the private sector by making it possible to employ
youngsters for a two-year trial period during which flexible redundancy
rules would apply (which meant, for example, that they could be laid
off without justification). The reaction to the law was guided by two concerns.
First, it was seen as an institutionalized form of job insecurity. Second,
Prime Minister Dominique de Villepin tried to push the law through
parliament by decree, without public debate and consultation with the unions.
Faced with massive street protests, the government eventually withdrew
the CPE.

6 The RPR (*Rassemblement pour la République*) was created in 1976 by Jacques
Chirac. It was later dissolved into the UMP in 2002 (*Union pour un Mouvement
Populaire*), which is currently presided over by Nicolas Sarkozy.

7 Pasqua was the Minister of the Interior in the Chirac government of 1986–8
as well, and had earned a reputation for his repressive measures.

8 A report published by the *Cour des Comptes* (Court of Accounts) in 2002 also
distinguished between two different geographies of priority neighbourhoods,
referring to them as 'contractual' (until mid-1990s) and 'statist'.

CHAPTER 3   THE RIGHT TO THE CITY? REVOLTS AND THE INITIATION OF URBAN POLICY

1 '*Autogestion*' is usually translated as 'self-management', and it means joint
management by workers and directors. I use the term without translating in
order to emphasize Lefebvre's particular conceptualization of *autogestion*,
which was not only industrial but urban as well. He tried, in a sense, to extend
the idea from industrial relations to the city and the production of its spaces,
a production that implied the appropriation of space by inhabitants. He used
the terms 'urban *autogestion*' (1970) and 'territorial *autogestion*' (1978), which
he conceived as a form of resistance to the 'administrative rationality' of the
state, or to the 'statist logic' (*la logique étatique*), through the appropriation of
space (Lefebvre, 1978: 323).

2 Pierre-Didier Tchétché-Apéa from Vaulx-en-Velin recalls: 'When we arrived,
everything was new. In a way, it was a social ascension, because everything
was clean and neat, each person had his/her room, it was luminous. The flats
are rather well done' (Tchétché-Apéa, 2000: 79).

3   This law was passed in a period of intense mobilization by second-generation immigrants of North African origin, epitomized by the highly publicized 'March for Equality and Against Racism' in 1983.

4   Cited in Collovald (2001: 104, fn. 6).

5   Cited in Feldblum (1999: 173, fn.13).

6   For an account of the revolts in Brixton, see Keith (1993).

7   This population loss raises a question about the degree of active increase in the unemployment rate. Even if one assumes that the part of the active population that left the commune in this period had remained, the unemployment rate in 1982 would still have been 10.1%, still more than the departmental (7.6%), regional (7.7%) and national (8.9%) rates.

8   As Bachmann and Le Guennec (1996: 353–4) show, the reception of what happened in several cities of Britain in the French media was conditioned by the 'particularity' of the French Republic, where, in contrast to Britain, race riots were not possible. This connects to the ideal of republicanism briefly discussed in Chapter 1, which emphasizes a common culture and identity.

9   *Commission Nationale pour le Développement Social des Quartiers*. Although the CNDSQ was created in 1981 and had immediately started to work in the following year, the official decree for its creation was not published until 1986. In the decree published in 1986, the CNDSQ's mission was defined as follows: 'The commission's mission, within the framework of State-Region contracts for the social development of neighbourhoods in difficulty, is to study and propose to the government all kind of actions likely to prevent physical and social degradation of these neighbourhoods' (Décret no. 86–183 du 6 février 1986 portant creation d'une Commission nationale pour le développement social des quartiers, published in the *Journal Officiel* on 8 February 1986, p. 2258).

10 This practice continued after Dubedout with Rodolphe Pesce (mayor of Valence), François Geindre (mayor of Hérouville Saint-Clair) and André Diligent (mayor of Roubaix), who served as presidents of the CNDSQ. It was, however, changed with the further institutionalization and centralisation of urban policy in 1988 through the creation of the *Délégation Interministérielle à la Ville* (DIV). Although the first president of the DIV was Yves Dauge, a mayor, its successive presidents have been, since 1991, either senior civil servants or prefects. This change is indicative of the shifting orientations of the French state in the 1990s, as I will try to show in the following chapters.

11 Dominique Figeat recalls: 'I remember, during the first meetings, the state civil servants were absolutely horrified at having to debate financial decisions, or administrative decisions, in front of other partners. They considered they shouldn't be involved in state decisions' (interview with the author).

12 The name of the programme – 'Social Development of Neighbourhoods' – was inspired by the work of Edgar Pisani on development. Compared to the previous Housing and Social Life (HVS) programme, it was seen as a more dynamic approach. 'As opposed to social life because "Housing and Social Life" was about housing and social *life*. So social development implied a dynamics, you see? [. . .] A social dynamics. A dynamics of economic and social development,

actually, because it wasn't necessarily just social. It wasn't just about repairing, it was about repairing and developing' (interview, Sylvie Harburger).

13 Another proposal advanced by the report was the creation of a national council for the prevention of delinquency, which was realized in 1983 (*Conseil National de Prévention de la Délinquance*, CNPD).

14 'Because at the time of the Dubedout Report, and around that time, we were carried by an ideology, too, that was very, you know . . . I don't know who was right. We used to say that's what the working class is like nowadays, they're mixed, ethnically. We took it as a fact, we weren't passing any judgement. We used to say that's how it is, nowadays the working class is coloured' (interview, Sylvie Harburger).

15 One is struck by the report's Lefebvrian tone. Pierre Saragoussi, one of the members of the commission, recalls: 'Well, this democratization, it wasn't just . . . political. It was, but it was also about participating in everyday life. And there we were, well, me personally, I was, steeped in Lefebvre's ideas, Henri Lefebvre, about everyday life, about, . . . all those things . . . well, I'd read, I'd *devoured*' (interview, Pierre Saragoussi).

16 Another programme with a spatial approach was conceived around the same time as DSQ: *Zones d'Education Prioritaires* (Educational Priority Areas, ZEPs). The aim of this programme, introduced in December 1981, was to provide additional resources, such as extra teachers, to schools in designated areas ('zones').

17 As Sylvie Harburger and Pierre Saragoussi explain, there were also political considerations, such as finding a balance between left and right municipalities, north and south, and east and west.

18 The label of the programme was carefully chosen, with an implication of the French Revolution in 1789. The year 1989 would also be the first year of the second term of Mitterrand, if re-elected, which he was.

19 The Socialist mayor of Dreux, Françoise Gaspard, was thus replaced with the coming to power of a right-wing party with close links to the FN, and Stirbois entered the municipal council. In a 1985 interview, he stated that the left had 'lit the fire' while trying to 'do revolution' with the immigrants, which his party was trying to 'extinguish'. 'Civil war will occur', he declared, 'if measures are not taken' (*Le Monde*, supplementary issue, 'Les immigrés: enjeu électoral', 24–5 November 1985: 3).

20 Two examples from Paris: Alain Juppé's (RPR, Prime Minister under Chirac between 1995 and 1997) electoral bulletin claimed a relationship between 'clandestine immigration, delinquency, and criminality'. A pamphlet signed by him and his fellow party members stated that 'this invasion [had] to be stopped'. The UDF candidate Jacques Dominati's electoral bulletin contained the message, 'Let us demand the right to security', printed under a photo showing black people (*Le Monde*, 13–14 March 1983: 9). The right-wing press also did its best. The weekly magazine *L'Express*, for example, published a special issue a month before the elections, the cover of which read, in bold capital letters, 'Immigrés: le dossier explosif' (28 January–3 February 1983).

21 The highly mediatized anti-racist organization *SOS-Racisme* was also created in this period, in 1984, with close financial links to the Socialist Party.

22 'For me, what you have behind it, it's two different ways of thinking about public action, that were in a sort of confrontation. And these two ways of thinking were quite well reflected in the two personalities, of Mitterrand, on the one hand, and of Dubedout, on the other. They weren't in the same category of politicians of course, but they knew each other and hated each other heartily. Mitterrand never wanted Dubedout to become minister, though he was, at the time he was considered, in the 1980s, as one of the great Socialist locally elected officials. [. . .] Mitterrand considered that Dubedout was not a real politician, he was someone from an associative background who took on political responsibilities, and, and Dubedout, I believe, hated Mitterrand's way of doing politics. [. . .] But Dubedout had a way of conceiving public action that emphasized people's everyday problems, actions associating citizens and associative movements, and for him political figures were, in a way, just the representatives of citizens, with a mandate to act on behalf of a collectivity. Well, needless to say, Mitterrand had a different way of seeing political action and public action' (interview, Dominique Figeat).

23 National Plans, prepared every five years, provide guidelines for public- and private-sector investment decisions.

24 The term 'cohabitation' is used when the President and the Prime Minister are from different parties or blocs (Mitterrand was from the PS, Chirac was from the RPR–UDF coalition). The 1986 cohabitation was the first time in France. Later, however, there were other cohabitations: Mitterrand (PS)–Balladur (RPR) between 1993 and 1995, and Chirac (RPR)–Jospin (PS) between 1997 and 2002.

25 Loi no. 88–1088 du 1er décembre 1988 relative au revenu minimum d'insertion (RMI). Published in the *Journal officiel* of 3 December 1988, p. 15119.

CHAPTER 4   JUSTICE, POLICE, STATISTICS: SURVEILLANCE OF SPACES OF INTERVENTION

1 Cited in Bachmann and Le Guennec (1996: 430). This 'notorious example', they state, owes much to the publication of the 1990 Sardais Report.

2 Mathieu Kassovitz's influential movie *La Haine* (Hate), released in 1995, was shot in the social housing neighbourhood of Noé in Chateloup-les-Vignes, where incidents had occurred. *La Haine* was about three young people (one Arab, one Jew, one black) in this *banlieue* of Paris, and the movie was dedicated to the young people who had died at the hands of the police, to 'those who died while this film was being made'. The script was published the same year with the title *Jusqu'ici tout va bien* (So far so good) – the tag line of the movie. A review of the movie may be found in Elstob (1997–8), who notes that the impact of the movie was so strong that the Prime Minister and the Minister of the Interior watched it three times in a private screening. See also Sharma and Sharma (2000) for a discussion of the movie.

3 The incidents had occurred in 1990.

4 A more detailed account of the incidents of Vaulx-en-Velin is presented in Chapter 6.

5   The headscarf affair of 1989 started with the expulsion of three female students from school for refusing to remove their Islamic headscarves, which was judged by the school's headmaster to be a breach of the French law on secularism (*laïcité*) in state schools. The then Minister of Education, Lionel Jospin, overturned the headmaster's decision. Another headscarf affair took place in 1994, this time upon the initiation of the Minister of Education, François Bayrou, who sought to ban 'ostentatious' signs of religion in public schools (which, as he admitted, included headscarves but not crucifixes or yarmulkas). In 2004, a law that prohibited 'ostentatious' signs of religion in public schools was finally passed. Shortly after the first headscarf affair, Prime Minister Rocard created the *Haut Conseil à l'Intégration* (High Council for Integration, the HCI). It should be remembered that the headscarf was a strong symbol of resistance to the colonial law. For more on the headscarf affairs and their implications, see, among others, Duchesne (2005); Hargreaves (1995); Hargreaves and McKinney (1997); Silverman (1992).

6   See Cesari (1997) for a detailed account of the 'fear of Islam'.

7   In an excellent analysis of the media construction of 'Islamphobia' in France, Deltombe (2005) shows that during the media hysteria around the Islamic headscarf affair that took place only a year before Vaulx-en-Velin, 'the *banlieue*' was totally absent from the debates. After the revolts, however, Islam and *banlieue* were immediately articulated together, never to be separated again.

8   For more on the 'Khaled affair', see Deltombe (2005) and Hargreaves and McKinney (1997).

9   It is not an uncommon police practice in France to perform identity checks on young people with darker complexion, or to retain those with darker complexion for questioning while letting others go. See, for example, Mucchielli (2001: 105–6).

10   Nakano (2000: 97–8) argues that the institutional origins of the prefect can be traced further back than the Napoleonic era, to the *intendant* of the kings, and that the change in title carries, in this sense, a revolutionary connotation. The prefect's title was restored when the right was in power again after the 1986 elections. For more on the history of this institution, concerned particularly with the maintenance of public order, see Ebel (1999).

11   The number of sous-prefects was raised to 15 the next year, and to 30 in 1993. By the end of the decade, there were 31 sous-prefects.

12   Most taxes in France are levied by the state, which then redistributes some of its revenue to local authorities in the form of what is called the *Dotation globale de fonctionnement*, which complements the revenue of local authorities when needed. Particularly deprived communes are entitled to additional funds, and this is where this new law intervenes. A form of solidarity payment is levied on the revenue of communes classified as rich, and then redistributed to communes classified as poor.

13   When the LOV was in preparation, a seminar was organized (by Véronique de Rudder) to start a dialogue between law-makers and researchers. The minutes of the seminar were then published in *Recherches* 20 (1991). De Rudder's remark appears on p. 36.

14 Tax concessions were also introduced for the first time, although this measure was never widely applied. As we will see in the next chapter, another urban policy programme in 1996 would put an extended version of this measure at the core of urban policy.

15 Loi no. 95–74 du 21 janvier 1995 relative à la diversité de l'habitat. Published in the *Journal officiel* of 24 January 1995, p. 1263. In terms of social housing, this law modified the LOV in three ways. First, it restricted the scope of the LOV by introducing a population threshold of 3,500 inhabitants (whereas the LOV required all communes in agglomerations of minimum 200,000 inhabitants). Second, it repealed the article about the possibility of using the right of preemption if no plan was prepared by the communes to meet the requirements of social housing provision. Finally, it repealed the articles about 'participation to housing diversity', which sought to create funds for social housing construction through charges from developers.

16 '*Sans foi ni loi*' is a harsh expression. It is usually translated into English as 'lawless', and its origins go back to medieval times, where the term was used to refer to bandits and outlaws, who obeyed neither the Church (*sans foi*) nor the Lord (*ni loi*), who had neither religion nor morals. It literally means 'faithless and lawless'. Faithless, no respect for or recognition of the law, and no morals: the expression '*sans foi ni loi*' also carries bestial connotations. In the year 2000, the expression would appear on the cover of *Les cahiers de la sécurité intérieure*, a journal published by the Institute for Advanced Studies of Interior Security, as the title of a special issue on the delinquency of young people (*jeunes*): *Jeunes: sans foi ni loi?* (no. 42). During the revolts of autumn 2005, Minister of the Interior Nicolas Sarkozy would also use the expression to refer to those involved in the incidents. See his article published in *Le Monde* on 6 November 2005, entitled 'Notre stratégie est la bonne'. The online version appeared on 5 November 2005 and can be accessed at http://www.lemonde.fr/web/article/0,1–0@2–3232,36–706906,0.html (last accessed 30 December 2006).

17 Hargreaves argues that the HCI, in theory, was concerned with integration in general; it was not conceived exclusively with the immigrant populations in mind. In practice, however, it 'focused almost exclusively on the population of immigrant origin' (1995: 196).

18 The title in original is *La justice agit dans la ville*. Here an explanation of the several meanings of French '*la justice*' seems necessary. The same word is used to refer both to justice (as an ideal, principle, or to refer simply to the penal system) and to law (although there is '*la loi*', which also means law). The use of the term in official documents (especially those prepared by the Ministry of Justice or the Ministry of the Interior) almost always implies the penal system and the power of imposing the rule of law. There is, indeed, a major difference between the French '*la justice*' and the English 'justice' in that the former implies 'the power of imposing the rule of law', and enforcement and sanction, which seem to be given less emphasis in the English definitions of the notion. See Weber (1992: 233–4) for the differences between the French and English definitions of 'justice'.

19 See, for example, the interview published in *Le Monde* (8 December 1998: 17), which presents Bui-Trong as 'an intellectual at the RG'.

20 The list of riots is kept by the 'Cities and *Banlieues*' section at the RG. The list, I was initially told, was not even released to the press. Later, however, Lucienne Bui-Trong kindly accepted to provide me with a copy of the list of large-scale riots (degree 8, according to her scale), which also includes information on the 'triggering incidents' and circumstances. A detailed analysis of these incidents, including the revolts of autumn 2005, is presented in Chapter 7.

21 A clarification as to the data used in the table should be made here. The results of this statistical profiling of the priority neighbourhoods were published by statisticians who worked in the 'City' section at INSEE. The data used by Champion et al. (1993) were based on 515 of the 546 neighbourhoods. Three years later, Champion and Marpsat (1996) published other findings based this time on 532 neighbourhoods. I use figures taken from the database created through a recent collaboration of DIV and INSEE (INSEE-DIV, no date), which provides information on the priority neighbourhoods of urban policy for 1990 and 1999. This database includes 751 priority neighbourhoods. It includes, therefore, information on both the 546 priority neighbourhoods in 1990 and the ones added later, without making a distinction. This creates a problem when presenting quantities, but not proportions. For example, there were 3 million inhabitants living in the priority neighbourhoods in 1990, although this number is given as 4.7 million in the recent database, an increase due to the increased number of neighbourhoods used in the calculation. I keep this problem in mind when presenting quantities. In terms of proportions, however, the data presented in the recent database do not diverge significantly from the data presented by the statisticians cited above. I therefore use this database, which provides information that is not offered in the cited articles above (except for nationalities, which are not included in the database).

22 More on the politics of this new statistical category will be presented in Chapter 7.

23 'Because you know, I saw Michel Rocard again recently! So, he was standing there, so I went to talk to him, said, you know, I worked for you, in a way, at a time. I was involved in urban policy. He said to me, "you know, now, looking back, I realize I was induced to make a mistake". It was precisely to shift from handling a few neighbourhoods, to one hundred, one hundred and forty, one hundred and fifty, [. . .] four hundred' (interview, Pierre Saragoussi).

24 '*Double peine*' literally means 'double punishment', and is one of the major issues the MIB has been fighting about (starting with the 'Resistance of *Banlieues*' period). Indeed, the term was conceived by them and has been rather successful as a rallying slogan. What is implied is prison plus expulsion: if a foreigner commits a crime that is judged to be punished by confinement, s/he first serves her/his term in the prison, and then is sent to her/his 'country of origin', even though that might not mean much for the person involved. In 2003, Minister of the Interior Nicolas Sarkozy claimed to have 'abolished'

double peine with a new law on immigration. However, as Lochak's (2004) analysis makes clear, it was far from abolished; indeed, it was expanded in many ways. See also the analysis in Gisti (2004).

25   The political engagement of the founding members of the MIB, however, goes as back as the early 1980s. More information can be found at: http://mibmib. free.fr/mib01.html (last accessed 31 December 2006).

26   As Silverman (1992: 6) argues, the 'demonization' of Le Pen and his party, *Front National*, by the anti-racist movements of the 1980s 'failed to locate the complex nature (and causes)' of racism in France.

27   The extreme right leader Le Pen's formulation was as follows: 'We not only have the right but the duty to defend our national personality, and we also have our right to difference' (cited in Feldblum, 1999: 36).

28   Pasqua was the Minister of the Interior in the Chirac government of 1986–8 as well, and had earned a reputation for his repressive measures.

CHAPTER 5   FROM 'NEIGHBOURHOODS IN DANGER' TO 'DANGEROUS NEIGHBOURHOODS': THE REPRESSIVE TURN IN URBAN POLICY

1   Pierre Méhaignerie was Minister of Justice in the Balladur government.

2   Circulaire du 31 octobre 1995. 'Renforcement de l'action de la Police Nationale dans la lutte contre les violences urbaines' (Ministère de l'Intérieur), p. 1.

3   Circulaire du 22 novembre 1994. 'Géographie des quartiers prioritaires de la politique de la ville au XIè Plan' (Le Ministre d'Etat, Ministre des Affaires Sociales, de la Santé et de la Ville), p. 2.

4   *'Pacte de Relance pour la Ville'* cannot be translated accurately, although it has been referred to as 'urban renewal pact' (see, for example, OECD, 1998). The word *'relance'* has economic connotations, and means 'boosting' when used with reference to the economy, and 'relaunching' when used with reference to a project or idea.

5   The project for the new law, the *Pacte de Relance pour la Ville*, was discussed at the Senate on 8 October 1996, with the opening statements of Jean-Claude Gaudin, Minister for Territorial Planning, City and Integration. The quotes are from the minutes of this discussion, which may be accessed via the following link: http://www.senat.fr/seances/s199610/s19961008/sc19961008001. html (last accessed 31 December 2006).

6   The quotes are from the minutes of the discussion of the law at the Senate. Gaudin is currently the mayor of Marseille, and has been known for his tough attitude towards the city's immigrant, notably Arab, population since he started office. Hargreaves (1995: 182), for example, wrote: '[I]n the final stages of the campaign for the municipal elections held in March 1983, Defferre and his centre-right opponent for the mayoralty of Marseille, Jean-Claude Gaudin, vied with each other in their claims as to who would be toughest in dealing with the city's Arab population.'

7   There was another addition to the 'characteristic elements of *cités*' in the 1990s: drugs. Eric Raoult, the then minister charged with integration and fighting

against exclusion, defined the 'two diseases of *cités*' as 'unemployment and drugs' (Sénat, 1995: 6). This 'characteristic' was immediately picked up by the media. When the *Pacte de relance* programme was announced, for example, *Le Monde* presented it at the top of its front page as the 'plan for the *banlieues*' with a cartoon that showed a street corner (with stereotypical images of buildings and young people that suggested that it was located in the *banlieue*) with shop names such as 'hashish to go', 'crack in stock' and 'good shots' (*Le Monde*, 19 January 1996:1).

8   From the minutes of the discussion of the *Pacte de Relance* at the Senate (see note 5 above).

9   Loi no. 95–115 du 4 février 1995 d'orientation pour l'aménagement et le développement du territoire.

10 The law was published in the *Journal officiel* of 15 November 1996. The definitions are taken from p. 16656.

11 This might give the impression that the urban policy neighbourhoods are areas where the majority of long-term unemployed people live. However, in the 1990s, 85% of the long-term unemployed in metropolitan France lived not in the urban policy neighbourhoods but elsewhere. The proportion of long-term unemployed living in the urban policy neighbourhoods remained 15% in both 1990 and 1999, although their number increased by 110,000 in the neighbourhoods and by 675,000 in metropolitan France. Calculated from INSEE (1990 and 1999a) and INSEE-DIV (no date).

12 The formula was as follows: ISE = (% of people younger than 25) × (% of long-term unemployed) × (% of people without high school diploma) × (total population of the commune) divided by the tax potential of the commune where the neighbourhood in question is located. The higher the ISE value, the more 'excluded' was the neighbourhood.

13 Décret no. 96–1154 du 26 décembre 1996 portant délimitation de zones franches urbaines dans certaines communes, available at: http://www.admi. net/jo/AVIV9604447D.html (last accessed 31 December 2006).

14 The programme was restricted to non-profit and public organizations only, in order to avoid a possible substitution of the existing personnel in the private sector by subsidized employees (Levy, 2005).

15 The MJDs were first conceived in 1991 (see Chapter 4). There were about 50 MJDs when the Sueur Report was published.

16 Circulaire du 28 octobre 1997 relative à la mise en œuvre des contrats locaux de sécurité (Ministère de l'Interieur). Published in the *Journal officiel* of 30 October 1997, pp. 15757–9.

17 The Council was created with a decree published on 19 November 1997 (Décret no. 97–1052 du 18 novembre 1997 créant le Conseil de sécurité intérieure).

18 The idea of creating a *police de proximité* was already there in the early 1990s when the Ministry for the City had started to collaborate with the Ministry of Justice (see Chapter 4). The then Minister of the Interior, Philippe Marchand, had introduced the idea at the Council of Ministers on 20 November 1991. The *police de proximité* meant more police on the beat, with a knowledge of the neighbourhood, closer to inhabitants, aimed more at prevention than

repression. The Jospin government tried to re-vitalize the measure, although it was somehow looked down upon and resisted in the police establishment. When Nicolas Sarkozy became Minister of the Interior in 2002, he mocked the measure by calling it 'police for the meek' (*'police des gentils'*), prioritized repression over prevention, and eventually dismantled this institution. For a detailed account of this measure, see Roché (2005), who argues that a veritable *police de proximité* focused on prevention could have averted the incidents of autumn 2005.

19 Jospin's speech is available at: http://www.archives.premier-ministre.gouv.fr/jospin_version2/PM/D270199.HTM (last accessed 31 December 2006). A more detailed list of the decisions made by the council is available at: http://www.archives.premier-ministre.gouv.fr/jospin_version2/PM/RELEVE.HTM (last accessed 31 December 2006).

20 The decisions made at the 19 April 1999 meeting of the *Conseil de Sécurité Intérieure* are available at: http://www.archives.premier-ministre.gouv.fr/jospin_version2/PM/RELEVE2.HTM (last accessed 31 December 2006).

21 A more detailed discussion of the CLS will be presented in Chapter 7 below.

22 The list of the MJDs may be found on the website of the Ministry of Justice via http://www.justice.gouv.fr/ville/mjd.htm (last accessed 31 December 2006). The number of MJDs reached 117 in 2006.

23 Circulaire du 31 décembre 1998 relative aux contrats de ville 2000–2006 (Premier ministre). Published in the *Journal officiel* of 15 January 1999, p. 726.

24 Loi no. 99–533 du 25 juin 1999 d'orientation pour l'aménagement et le développement durable du territoire et portant modification de la loi no 95–115 du 4 février 1995 d'orientation pour l'aménagement et le développement du territoire. Published in the *Journal officiel* of 29 June 1999, p. 9515.

25 Loi no. 99–586 du 12 juillet 1999 relative au renforcement et à la simplification de la coopération intercommunale. Published in the *Journal officiel* of 13 July 1999, p. 10361.

26 Loi no. 2000–1208 du 13 décembre 2000 relative à la solidarité et au renouvellement urbains. Published in the *Journal officiel* of 14 Decembre 2000, p. 19777.

27 'For instance, one day I went to an area where we had a social housing project. The inhabitants of the neighbourhood didn't want to hear of it, because, in fact, they didn't want any immigrants in their neighbourhood. Well, I'm sorry, I didn't listen to them, and I built social housing' (interview, Maurice Charrier).

28 Circulaire du 31 décembre 1998 relative aux contrats de ville 2000–2006 (Premier ministre). Published in the *Journal officiel* of 15 January 1999, p. 726.

29 The figures below are taken from the database created through the collaboration of DIV and INSEE (INSEE-DIV, no date). There were some missing sites in the file, and the ZFU of Marseille was among them.

30 The active increase in the unemployment rate in these neighbourhoods may be less intense than it seems at first sight, for they have been losing population as well. As the Sueur Report (1998) quite explicitly put it, nobody wants to live in a 'zone'. Therefore, one could suppose that inhabitants who can afford to do so, the ones with jobs presumably, leave these neighbourhoods, which makes the increase in unemployment rates look more than it actually is. The rate of population decrease in the priority neighbourhoods of urban policy from 1990 to 1999 was 5.7%. Even if we assume that all the people left were part of the active population, and re-calculate the level of unemployment accordingly, the figure would be 24.6%, not diverging greatly from the actual figure, which is 25.4%.

31 Loi no. 2001–1062 du 15 novembre 2001 relative à la sécurité quotidienne ('loi LSQ'). Published in the *Journal officiel* of 16 November 2001, p. 18215.

32 Loi no. 2002–1094 du 29 août 2002 d'orientation et de programmation pour la sécurité intérieure. Published in the *Journal officiel* of 30 August 2002, p. 14398. Loi no. 2003–239 du 18 mars 2003 pour la sécurité intérieure ('loi LSI'). Published in the *Journal officiel* of 19 March 2003, p. 4761.

33 Loi no. 2002–1138 du 9 septembre 2002 d'orientation et de programmation pour la justice ('loi Perben I'). Published in the *Journal officiel* of 10 September 2002, p. 14934. Loi no. 2004–204 du 9 mars 2004 portant adaptation de la justice aux évolutions de la criminalité ('loi Perben II'). Published in the *Journal officiel* of 10 March 2004, p. 4567.

34 With these laws, not only was more power given to the police, but also new infractions were introduced. This meant increased work for the juridical system. To this end, another measure, called 'proximity judges' (*'juges de proximité'*), was introduced in 2003 to hire 3,300 people (compared to only 180 qualified magistrates) to deal with smaller infractions. The measure was highly criticized by the Magistrates Union, which argued that the so-called 'judges' would exercise the functions of a magistrate without really being trained. Therefore, ordinary people would be judged without juridical guarantee by judges who were not magistrates (they would, nevertheless follow a five-day training programme at the *Ecole Nationale de la Magistrature*), hired only for a non-renewable period of seven years. Furthermore, their recruitment would be done locally, the Union held, opening the way to pressures and manipulations. The Union's statement is available from the author (original web link no longer functioning). The reference for the law creating the proximity judges is: Loi no. 2003–153 du 26 février 2003 relative aux juges de proximité. Published in the *Journal officiel* of 27 February 2003, p. 3479.

35 '[W]hat strikes me is that political life [in France] is organized around Le Pen', boasted the extreme right leader in December 2005. And not without reason: since 2002, 16 out of 30 propositions of the extreme right's 'Justice and Police' programme have been realized or are in the course of being realized (*Le Canard enchaîné*, 14 December 2005: 4).

36 That said, in 2001 the State Secretary for Housing, Marie-Noëlle Lienemann (PS), was already talking about 'breaking up the ghettos' with reference to the renewal programmes of the Jospin government (*Libération*, 2 October 2001:

21). However, the Jospin government did not prioritize this as an objective, and was more careful in its use of the term. In 2001, when the demolition and renewal programmes intensified, the stated aim was to 'turn the page of dormitory towns', although occasional references to ghettos were made. It was held, for example, that urban policy had not prevented the 'process of ghettoization of the popular neighbourhoods of certain cities' (CIV, 2001a: 3).

37 Loi no. 2003–710 du 1er août 2003 d'orientation et de programmation pour la ville et la rénovation urbaine. Published in the *Journal officiel* of 2 August 2003, p. 13281.

38 However, the law kept open the possibility of including 'exceptional cases', which showed similar characteristics to the ZUSs but were not designated as such.

39 This again can be seen to follow an idea already present in the *Pacte de Relance*, which required an annual report on the ZFUs to be presented at the parliament.

CHAPTER 6   A 'THIRST FOR CITIZENSHIP': VOICES FROM A *BANLIEUE*

1 Document number 030313, dated 12 November 1991. The official name of the association is '*Agora Vaudais*' (*Vaudais* meaning of/from Vaulx-en-Velin). However, it is commonly known as *Agora*.

2 Curiously, there was no high school in Vaulx-en-Velin, a city of about 45,000, before the incidents. Although the city had been asking for one for about twenty years, establishing a high school in Vaulx-en-Velin was not considered necessary by the regional authorities. 'And three months after the riots, yes, it was necessary to establish a high school' (interview, Yves Mena).

3 For a more detailed account, see Chabanet (1999: 358, fn. 3).

4 When asked about this accusation, the mayor did not object to it, and stated that some, but certainly not all, the members of *Agora* were 'in the logic of communitarianism'.

5 A common practice in the preparation of lists by the major political parties is to put someone from 'immigrant origin', notably North African, towards the end of their list to show their 'openness'. This is referred to as '*Arabe de service*' – 'token Arab'.

6 A similar remark was made by Pierre-Didier in an interview given to a local newspaper: 'The people of the right are clearer. They say to us "I don't like you, and this is what I have in store for you". The left says to us "I love you", and it plays revolting tricks on us from behind.' A copy of the interview was obtained from the archives of the association. Although there is no reference on it, the newspaper is probably *Le Progrès de Lyon*, and the year 1994. A copy of the interview is available from the author.

7 Here it is important to note the context in which the interviews were conducted. It was shortly after the 2002 presidential elections, where the issue of security seemed decisive. Indeed, there was little difference in this sense between the campaigns of the right, extreme right and mainstream left parties.

In the social housing neighbourhoods of *banlieues*, in Vaulx-en-Velin as in others, tension was building up as the then new Minister of the Interior, Nicolas Sarkozy, had announced his security measures aimed mainly at the *banlieue* youth, which included easier identity checks and provision of flash-ball guns to police officers working in such neighbourhoods.

8 This call is available via http://lmsi.net/article.php3?id_article=336 (last accessed 1 January 2007).

9 *SOS-Racisme* has strong financial links with the Socialist Party, and its successive presidents eventually find their ways to the party. Monitoring discriminatory practices in the housing market, holiday establishments and night-clubs constitute some of the most publicized activities of *SOS-Racisme*.

10 This trend – high abstention rates and support for the National Front – continued in the 2002 presidential elections. The abstention rates were 35.6% and 25.2% for the first and second rounds. In the first round, National Front leader Le Pen obtained the highest score in the commune with 21.7% of the votes. He was followed by Socialist Party leader Jospin (18.5%) and Chirac (12.0%). The Communist Party candidate Robert Hue obtained 6.1% of the votes. In the second round, Le Pen obtained a score of 22.4% and his rival, Chirac, 77.6%.

CHAPTER 7   VOICES INTO NOISES: REVOLTS AS UNARTICULATED JUSTICE MOVEMENTS

1 Although, as noted in Chapter 5, the Minister of the Interior, Nicolas Sarkozy, insisted on the 'perfectly organized' nature of the revolts, this claim was contradicted by a report by the French Intelligence Service, which interpreted them as a 'non-organized insurrection', a 'popular revolt of *cités*' (*Libération*, 8 December 2005: 16).

2 Literally 'I got the hate', this phrase is shorthand for the rebellion of the *banlieue* youth, as illustrated in the movie *La Haine* (Hate) by Mathieu Kassovitz (1995).

3 The information presented on the revolts of the 1990s is based on the list provided by Lucienne Bui-Trong, creator (in 1991) and head (until her retirement in 2002) of the 'Cities and *Banlieues*' section at the *Renseignements Généraux* (French Intelligence Service, RG).

4 Some communes experienced revolts more than once: Vaulx-en-Velin (1990 and 1992), Garges-lès-Gonesse (1991, twice in 1994 and 1995), Amiens (1991 and 1994), Tourcoing (1992 and 1993), 18th district of Paris (1993 and 1999), Bron (1993 and 1994), Grigny (1995 and 1999) and La Seyne-sur-Mer (1997 and 2000).

5 These latter communes are Vauvert in the department of Gard, and Montauban in the department of Tarn-et-Garonne. The incidents in these communes occurred in 1999. Montauban signed a demolition–reconstruction convention with ANRU after 2003.

6 The neighbourhood of Etats-Unis is still a priority neighbourhood of urban policy, after more than two decades since its inclusion in 1984.

7 Lagrange and Oberti's analysis is based on the first 62 ANRU conventions signed in July 2005. As of March 2006, this number was 108, with 165 waiting to be signed (CIV, 2006).

8 Not that the government invested heavily in urban policy. As the daily *Libération* wrote: 'Brutal cuts in youth employment scheme [*emplois-jeunes*], severe cuts in subsidies, disappearance of neighbourhood associations . . . In terms of urban policy, the right-wing governments since 2002 have been a disaster.' Indeed, Prime Minister de Villepin explicitly admitted the government's responsibility in the decline of the associative sector, which saw its funds disappear after 2002 (*Libération*, 8 November 2005).

9 When asked about the *Maisons de Justice*, Pierre-Didier responded as follows: 'The *Maisons de Justice* . . . Well, it's . . . in the beginning we found that interesting, because we thought that it was a way, not only to have justice better known, but also to have a local mediation. And that was, is interesting! But it doesn't work that way at all. So, today, there's only, only repression has any effect' (interview, Pierre-Didier Tchétché-Apéa).

10 In January 2002, the Human Rights League, the Union of Magistrates and the Union of Attorneys of France formed a commission of inquiry to address this issue of spatially 'targeted' police repression.

11 This commonly used phrase – originally coined by Victor Hugo – is one of the slogans of the MIB (*Mouvement de l'Immigration et des Banlieues*).

12 My account is based on Bachmann and Le Guennec (1996: 445–6), Cyran (2003: 30), and Jazouli (1992: 153–4).

13 The same issue was highlighted again on Amnesty International's annual report in 2006. I still have not obtained a copy of this report, but a brief newspaper article on it shows that the same problems are emphasized for the case of France – police violence, racist police attitudes against Muslims and 'minorities', and police impunity – which the report relates to the revolts of autumn 2005 in the *banlieues* (*Le Canard enchaîné*, 24 May 2006: 5).

14 No immediate official explanation was made. More than a week later, it was stated, following an investigation, that the mosque was not directly targeted by the police (*Le Monde*, 10 November 2005: 12).

15 La Duchère is a social housing neighbourhood (80% of the total housing stock) towards the north-western limits of Lyon's 9th district, included in urban policy since 1984. It was designated as a demolition–reconstruction site in 2005, and as a ZFU in 2006. Large-scale revolts took place in the neighbourhood in 1997 following the killing of a young person in a police station.

16 This scene was broadcast on TF1 on 6 November 2005. Here I use the transcription from a letter by Jean-Pierre Dubois, president of the Human Rights League, addressed to Minister of the Interior Nicolas Sarkozy. The letter is available on the website of association IPAM in a document on the revolts of 2005, entitled 'Le soulèvement populaire dans les banlieues françaises d'octobre–novembre 2005' (http://www.reseau-ipam.org/article.php3?id_article = 1147; last accessed 4 January 2007).

17 According to research conducted by Sofres-Cecodip, which analysed the subjects covered by the media between January and March 2002, people listening to the radio or television were exposed to the issue of insecurity three times

more than the issue of employment. And if the analysis is limited to the television only, the issue of employment had virtually disappeared (cited in *Claris*, 2002: 3).

18 I am grateful to director Richard Vargas and Coup d'œil Productions for providing me with a copy and the transcription of this documentary.

19 Note de service no. 14/03 du 24 février 2003 Direction départementale de la sécurité publique des Alpes-Maritimes, District de Cannes, Circonscription d'Antibes-Vallauris. A copy of the letter is published in *Le Canard enchaîné* (23 April 2003: 4).

20 A copy of the letter was published again by *Le Canard enchaîné* on (29 October 2003).

21 The occasion was the visit of the Minister following the killing of an 11-year-old boy, who apparently was caught in crossfire.

22 One of Sarkozy's fellow party members, Jean-Paul Garaud (UMP), announced that he would propose a new law giving the courts the possibility to 'withdraw French nationality' from naturalized French citizens 'who participate in urban guerrilla actions' (*Le Monde*, 11 November 2005: 12).

23 'The phrase "urban violence", so common nowadays, does not in fact account for all violence inflicted by any category of population in urban areas, but refers specifically to violence committed by young people, mainly against institutions, amongst which the police, in or close to working-class areas, politely called "sensitive" or "difficult" areas, which sociologically illustrates the notion of "dangerous classes", currently fixated on the impoverished areas of urban peripheries (of the outskirts of cities)' (Esterle-Hedibel, 2002: 377).

24 These figures are used not only by the media, but also by what Mucchielli (2001) calls the 'experts' of security: that is, owners or associates of private security companies that present themselves as experts in the media and political circles, and 'sell security' after having convinced people – especially mayors – that there is indeed a serious insecurity problem. So, for example, a former police chief and former adviser to the DIV published a book in which he argued that in six years, between 1992 and 1997, 'the volume of urban violence has virtually increased fivefold'. Two business associates, one of them president of a prospering private security company, published, the same year, a book with the argument that it had increased by 400% (both cited by Mucchielli, 2001: 58). See Mucchielli (2001) for a discussion, and for more information on such security 'experts'.

CHAPTER 8   CONCLUSION: SPACE, POLITICS AND URBAN POLICY

1 This law was invoked only twice before, for the war in Algeria and for the incidents in France's overseas territory New Caledonia in 1985.

References

LIST OF INTERVIEWS

Brun, Bernard, project director for the *Grand Projet de Ville* (GPV) of Vaulx-en-Velin, 2 May 2002, Vaulx-en-Velin.

Bui-Trong, Lucienne, creator and former head of the 'Cities and *Banlieues*' section at the *Renseignements Généraux* (French Intelligence Service, RG), 6 June 2002, Paris.

Charrier, Maurice, mayor of Vaulx-en-Velin, 3 July 2002, Gare de Lyon, Paris.

Della, Abdel, *Association Agora*, 23 May 2002, Mas du Taureau, Vaulx-en-Velin.

Farid, *Mouvement de l'Immigration et des Banlieues* (Immigration and *Banlieues* Movement, MIB), 21 December 2004, Paris.

Figeat, Dominique, former member of the *Commission Nationale pour le Développement Social des Quartiers* (National Commission for the Social Development of Neighbourhoods, CNDSQ), 1 August 2006, Paris.

Guy, *Mouvement de l'Immigration et des Banlieues* (Immigration and *Banlieues* Movement, MIB), 21 December 2004, Paris.

Harburger, Sylvie, former member of the *Commission Nationale pour le Développement Social des Quartiers* (National Commission for the Social Development of Neighbourhoods, CNDSQ), 4 July 2006, Brussels.

Lahlou, *Mouvement de l'Immigration et des Banlieues* (Immigration and *Banlieues* Movement, MIB), 21 January 2005, Paris.

Mena, Yves, *Association Agora*, 23 May 2002, 23 May 2002, Mas du Taureau, Vaulx-en-Velin.

Nordine, *Mouvement de l'Immigration et des Banlieues* (Immigration and *Banlieues* Movement, MIB), 24 March 2005, Paris.

Pirot, Jean, project manager (*chef de projet*) for social development in the framework of the City Contract of Vaulx-en-Velin, 23 May 2002, Vaulx-en-Velin.

Saragoussi, Pierre, former member of the *Commission Nationale pour le Développement Social des Quartiers* (National Commission for the Social Development of Neighbourhoods, CNDSQ), 18 July 2006, Paris.

Tchétché-Apéa, Pierre-Didier, president of *Association Agora*, 26 April 2002, Lyon, and 23 May 2002, Mas du Taureau, Vaulx-en-Velin.

PRIMARY SOURCES (IN CHRONOLOGICAL ORDER)

Reports, Circulars, Laws and Decrees

Peyrefitte A., 1977. *Réponses à la violence*. Rapport du comité d'études sur la violence, la criminalité et la délinquance, 2 vols (La Documentation française, Paris)

Schwartz B., 1981. *L'insertion professionnelle et sociale des jeunes*. Rapport au Premier ministre (La Documentation française, Paris)

Bonnemaison G., 1982. *Face à la délinquance: prévention, repression, solidarité*. Rapport au Premier ministre de la Commission des maires sur la sécurité (La Documentation française, Paris)

Dubedout H., 1983. *Ensemble, refaire la ville*. Rapport au Premier ministre du Président de la Commission nationale pour le développement social des quartiers (La Documentation française, Paris)

Pesce, R., 1984. *Développement social des quartiers. Bilans et perspectives 1981–1984*. Rapport au Premier ministre du Président de la Commission nationale pour le développement social des quartiers (La Documentation française, Paris)

Décret no. 84-531 du 16 juin 1984 portant création du comité interministériel pour les villes

Décret no. 86-183 du 6 février 1986 portant création d'une Commission nationale pour le développement social des quartiers. Published in the *Journal officiel* of 8 February 1986, p. 2558

Décret no. 88-738 du 3 juin 1988 relatif aux attributions du ministre d'Etat, ministre de l'équipement et du logement

Loi no. 88-1088 du 1er décembre 1988 relative au revenue minimum d'insertion (RMI). Published in the *Journal officiel* of 3 December 1988, p. 15119.

Levy, F., 1988. *Bilan/Perspectives des contrats de plan de développement social des quartiers*. Commissariat général du Plan (La Documentation française, Paris)

Sardais, C., 1990. *Rapport sur la mise en œuvre de la politique de la ville*. Inspection générale des finances (Paris)

Mitterrand, F., 1990. Discours prononcé par Monsieur François Mitterrand Président de la République devant les assises de Banlieue 89, Bron, le mardi 4 décembre 1990 (Présidence de la République, Service de Presse)

Circulaire du 21 février 1990 'Politique des villes et du développement social urbain en 1990' (Premier ministre)

Loi no. 90-449 du 31 mai 1990 visant à la mise en œuvre du droit au logement (loi Besson). Published in the *Journal officiel* of 2 June 1990, pp. 6551–4

Ville et DATAR, 1990. *148 quartiers: bilan des contrats de développement social des quartiers du IXe plan 1984/1988* (Paris)

CNV, 1990a. *Les casseurs du 12 novembre* (Paris)

CNV, 1990b. *Rapport de la mission du Conseil National des Villes à Vaulx-en-Velin*. Premier ministre (Paris)

Projet de loi d'orientation pour la ville, 24 avril 1991, Assemblée Nationale no. 2009

Conseil économique et social, 1991. *Avis sur le projet de loi d'orientation pour la ville*

Loi no. 91-662 du 13 juillet 1991. *Loi d'orientation pour la ville* (LOV). Published in the *Journal officiel* of 19 July 1991, pp. 9521–30

Circulaire du 31 juillet 1991. 'Circulaire no. 91-57 relative à la loi d'orientation pour la ville no. 91-662 du 13 juillet 1991' (Ministère de l'Equipement, du Logement, des Transports et de l'Espace, Direction de l'Architecture et de l'Urbanisme, Direction de la Construction)

Delarue, J.-M., 1991. *Banlieues en difficultés: la relégation*. Rapport au ministre d'Etat, ministre de la Ville et de l'Aménagement du territoire (Paris)

CNV, 1991a. *Mouvements collectifs et violence*, Pierre Cardo (Paris)

CNV, 1991b. *Media & quartiers* (Paris)

CNV, 1991c. *Participation des habitants à la ville*, Jacques Floch (Paris)

Ministère de la Justice, 1991. *La justice agit dans la ville* (Paris)

Circulaire no. 92-13 du 2 octobre 1992. 'Les réponses à la délinquance urbaine' (Garde des sceaux, Ministre de la justice)

CNV, 1992a. *Bilan des 64 propositions du rapport de la Commission des maires sur la sécurité 1982–1992* (Paris)

CNV, 1992b. *Analyse des résultats electoraux dans 20 quartiers en développement social* (Paris)

Dray, J., 1992. *La violence des jeunes dans les banlieues*. Rapport d'information par la Commission des affaires culturelles, familiales et sociales, Assemblée Nationale, no. 2832, 25 juin 1992

Haut Conseil à l'Intégration (HCI), 1992. *Conditions juridiques et culturelles de l'intégration* (La Documentation française, Paris)

Geindre, F., 1993. *Villes, démocratie, solidarité: le pari d'une politique*. Rapport du groupe Villes, Commissariat Général du Plan (La Documentation française, Paris)

Belorgey, J.-M., 1993. *Evaluer les politiques de la ville*. Comité d'évaluation de la politique de la ville (Paris)

Donzelot, J. and Estèbe, P., 1993 *L'Etat de solidarité*. Evaluation de la politique de la ville, vol. I (Délégation Interministérielle à la Ville, PLAN URBAIN, Caisse des Dépôts et Consignations, Paris)

Note de service, 31 mai 1994. 'L'action de la Sécurité Publique en matière de Police de Proximité (Ministère de l'Intérieur et de l'Amenagement du Territoire)

Circulaire du 22 novembre 1994. 'Géographie des quartiers prioritaires de la politique de la ville au XIè Plan' (Le Ministre d'Etat, Ministre des Affaires Sociales, de la Santé et de la Ville)

Loi no. 95-74 du 21 janvier 1995 relative à la diversité de l'habitat (loi Carrez). Published in the *Journal officiel* of 24 January 1995, p. 1263

Loi no. 95-115 du 4 février 1995 pour l'aménagement et le développement du territoire. Published in the *Journal officiel* of 15 November 1996, p. 16556

Décret no. 95-782 du 12 juin 1995 relatif aux attributions du secrétaire d'Etat aux quartiers en difficulté

Sénat, Commission des Affaires Sociales, 1995. 'Audition de M. Eric Raoult, Ministre chargé de l'intégration et de la lutte contre l'exclusion le jeudi 22 juin 1995' Assemblée Nationale compte rendu no. 6, Mardi 27 juin 1995

Programme National d'Intégration Urbaine, 1995. Rapport de présentation by Francis Idrac and Jean-Pierre Duport, le 8 septembre 1995 (Délégation Interministérielle à la Ville)

Circulaire du 31 octobre 1995. 'Renforcement de l'action de la Police Nationale dans la lutte contre les violences urbaines' (Ministère de l'Intérieur)

Circulaire du 20 novembre 1995. 'Mise en œuvre et évaluation des contrats de ville' (Le Ministre de l'Aménagement du Territoire, de la Ville, et de l'Intégration, Le Ministre délégué, chargé de la Ville et de l'Intégration)

Sénat, 1996. Rapport relatif à la mise en œuvre du pacte de relance pour la ville, presented by Gérard Larcher, 2 vols, Sénat, session ordinaire de 1996–7, no. 1

Loi no. 96-987 du 14 novembre 1996 relative à la mise en œuvre du pacte de relance pour la ville. Published in the *Journal officiel* of 15 November 1996, p. 16656

Circulaire du 17 juin 1996. 'Mise en œuvre du Pacte de Relance pour la Ville, Désignation des sous-préfets spécifiquement chargés de mission pour la politique de la Ville dans les départements où sont mis en œuvre des contrats de ville, Désignation de délégués de l'Etat dans les quartiers prioritaires' (Ministère de l'Aménagement du Territoire, de la Ville, et de l'Intégration, Ministère délégué à la Ville et de à l'Intégration, Ministère de l'Intérieur)

Décret no. 96-1154 du 26 décembre 1996 portant délimitation de zones franches urbaines dans certaines communes, available at: http://www.admi.net/jo/AVIV9604447D.html (last accessed 31 December 2006)

Conseil d'Etat 1996. *Rapport sur le principe d'égalité* (La Documentation française, Paris)

Delevoye, J.-P., 1997. *Cohésion sociale et territoires.* Commissariat Général du Plan (La Documentation française, Paris)

CNV, 1997. *Rapport du Conseil national des villes 1994–1997* (La Documentation française, Paris)

Circulaire du 15 mars 1997. 'Pacte de Relance pour la Ville, prévention de la délinquance en milieu urbain' (Délégation Interministérielle à la Ville, Paris)

Ministère de l'Intérieur, 1997. *Des villes sûres pour des citoyens libres. Les actes du colloque* (Ministère de l'Intérieur, Service de l'information et des relations publiques, Paris)

Chevènement, J.-P., 1997. 'Discours d'ouverture de Jean-Pierre Chevènement, ministre de l'Intérieur', in *Des villes sûres pour des citoyens libres. Les actes du colloque* (Ministère de l'Intérieur, Service de l'information et des relations publiques, Paris), pp. 3–14

Jospin, L., 1997. 'Discours de clôture de Lionel Jospin, Premier ministre', in *Des villes sûres pour des citoyens libres. Les actes du colloque* (Ministère de l'Intérieur, Service de l'information et des relations publiques, Paris), pp. 88–94

Circulaire du 28 octobre 1997 relative à la mise en œuvre des contrats locaux de sécurité (Ministère de l'Intérieur). Published in the *Journal officiel* of 30 October 1997, pp. 15757–9

Décret no. 97-1052 du 18 novembre 1997 créant le Conseil de sécurité intérieure

Circulaire du 31 décembre 1998 relative aux contrats de ville 2000–2006 (Premier ministre). Published in the *Journal officiel* of 15 January 1999, p. 726

CIV, 1998a. *Comité interministériel des villes, Dossier de presse*, 30 juin 1998, available at: http://www.ville.gouv.fr/infos/actualite/presse/civ0.html (last accessed 31 December 2006)

CIV, 1998b. *Comité interministériel des villes, Dossier de presse*, 2 décembre 1998, available at: http://www.ville.gouv.fr/infos/actualite/presse/civ.html (last accessed 31 December 2006)

Sueur, J.-P., 1998. *Demain, la Ville*. Rapport présenté au ministre de l'emploi et de la solidarité, 2 vols (La Documentation française, Paris)

Conseil de sécurité intérieure, 1999. *Intervention de Monsieur Lionel Jospin, Premier ministre, lors de la conférence de presse à l'issue du Conseil de sécurité intérieure*, 27 janvier 1999, available at: http://www.archives.premier-ministre.gouv.fr/jospin_version2/PM/D270199.HTM (last accessed 2 January 2007)

Conseil de sécurité intérieure, 1999. *Relevé de décisions*, 27 janvier 1999, available at: http://www.archives.premier-ministre.gouv.fr/jospin_version2/PM/RELEVE. HTM (last accessed 2 January 2007)

Conseil de sécurité intérieure, 1999. *Relevé de décisions*, 19 avril 1999, available at: http://www.archives.premier-ministre.gouv.fr/jospin_version2/PM/RELEVE2. HTM (last accessed 2 January 2007)

Loi no. 99-533 du 25 juin 1999 d'orientation pour l'aménagement et le développement durable du territoire et portant modification de la loi no 95-115 du 4 février 1995 d'orientation pour l'aménagement et le développement du territoire. Published in the *Journal officiel* of 29 June 1999, p. 9515

Loi no. 99-586 du 12 juillet 1999 relative au renforcement et à la simplification de la coopération intercommunale. Published in the *Journal officiel* of 13 July 1999, p. 10361

CIV, 1999. *Pour des villes renouvelées et solidaires* (Comité interministériel des villes du 14 décembre 1999, Paris)

DIV, 1999. *Chronologie des dispositifs de la politique de la ville* (Paris)

Loi no. 2000-1208 du 13 décembre 2000. *Loi relative à la solidarité et au renouvellement urbains* (loi SRU). Published in the *Journal officiel* of 14 December 2000, p. 1977

DIV, 2000a. *50 GPV: les grands projets de ville* (Les éditions de la DIV, Paris)

DIV, 2000b. *La reforme du dispositif des zones franches urbaines* (Les éditions de la DIV, Paris)

CNV, 2001. *Rapport du Conseil national des villes 1998–2001* (La Documentation française, Paris)

CIV, 2001a. *Comité interministériel des villes, Dossier de presse*, 1er octobre 2001 (Ministère délégué à la ville, Paris)

CIV, 2001b. *Comité interministériel des villes, le discours de Claude Bartolone, ministre délégué à la Ville*, 1er octobre 2001 (web link no longer functioning; copy available from the author)

CIV, 2001c. *Comité interministériel des villes, Discours à l'occasion du quatrième Comité interministériel des villes, Lionel Jospin, Premier ministre*, 1er octobre 2001, available at: http://www.archives.premier-ministre.gouv.fr/jospin_version3/fr/ie4/contenu/ 28922.htm (last accessed 2 January 2007)

Loi no. 2001-1062 du 15 novembre 2001 relative à la sécurité quotidienne ('loi LSQ'). Published in the *Journal officiel* of 16 November 2001, p. 18215

Anon., 2001. *Loi solidarité et renouvellement urbains. Des nouveaux outils pour les collectivités locales* (Direction générale de l'Urbanisme, de l'Habitat et de la Construction, Paris)

Cour des Comptes, 2002. *La politique de la ville*. Rapport public particulier (Les éditions des Journaux Officiels, Paris)

Loi no. 2002-1094 du 29 août 2002 d'orientation et de programmation pour la sécurité intérieure. Published in the *Journal officiel* of 30 August 2002, p. 14398

Loi no. 2002-1138 du 9 septembre 2002 d'orientation et de programmation pour la justice ('loi Perben I'). Published in the *Journal officiel* of 10 September 2002, p. 14934.

Loi no. 2003-153 du 26 février 2003 relative aux juges de proximité. Published in the *Journal officiel* of 27 February 2003, p. 3479

Loi no. 2003-239 du 18 mars 2003 pour la sécurité intérieure ('loi LSI'). Published in the *Journal officiel* of 19 March 2003, p. 4761

Loi no. 2003-710 du 1er août 2003 d'orientation et de programmation pour la ville et la rénovation urbaine. Published in the *Journal officiel* of 2 August 2003, p. 13281

DIV, 2004. *Chronologie des dispositifs de la politique de la ville* (Paris)

Loi no. 2004-204 du 9 mars 2004 portant adaptation de la justice aux évolutions de la criminalité ('loi Perben II') Published in the *Journal officiel* of 10 March 2004, p. 4567

CIV, 2006. *Pour une politique de la ville renouvelée* (Comité interministériel des villes du 9 mars 2006, Paris)

Statistics

INSEE, 1990. *Recensement général de la population, 1990* (INSEE, Paris)

INSEE, 1999a. *Recensement général de la population, 1999* (INSEE, Paris)

INSEE, 1999b. *Villes et quartiers sensibles face à la montée de la précarité* (INSEE, Rhône-Alpes)

INSEE-DIV, no date. *Fiches Profil – Quartiers de la politique de la ville: Données des recensements de la population de 1990 et 1999* (CD-ROM)

Newspaper Articles

L'Express, 1973. 'Banlieues: les "loubars" vous parlent', 3–9 September

Le Monde, 1981. 'Brixton en France?', 17 April, p. 11

Le Figaro, 1981. 'Marseille-Nord sous haute tension', 7 July, p. 26

L'Express, 1983. 'Immigrés: le dossier explosif', 28 January–3 February

Le Monde, 1983. 'Les urnes de la peur', 12 March, p. 10

Le Monde, 1983. 'Cher Mustapha . . .', 13–14 March, pp. 1 and 9

Le Figaro, 1983. 'François Mitterrand aux Minguettes', 11 August, p. 7

Le Monde, 1983. 'M. Mitterrand dans les ZUP de Saint-Etienne et de Vénissieux', 12 August, p. 6

Le Monde, 1983. 'Des beurs à l'Elysée, 4–5 December, pp. 1 and 11

Le Monde, 1983. 'Le succès et les conséquences du rassemblement contre le racisme', 6 December, p. 12

Le Monde, 1985. 'Les immigrés: enjeu électoral', supplement, 24–5 November

Le Monde, 1989. 'Un langage commun pour la sécurité', and 'La guerre est finie', 5–6 November, p. 8

Le Monde, 1990. 'Une radioscopie des "peurs" des Français', 4 July, p. 9

Le Figaro, 1990. 'Ces banlieues où le pire est possible', 9 October, p. 11

Le Monde, 1990. 'La mort d'un jeune motard provoque une émeute à Vaulx-en-Velin', 9 October, p. 13

Le Monde, 1990. 'Les vice-présidents du Conseil national des villes se rendent à Vaulx-en-Velin à la demande du premier ministre', 10 October, p. 14

Le Monde, 1990. 'Deux témoins contestent la version policière de l'accident de moto qui a provoqué la colère des jeunes de Vaulx-en-Velin', 11 October, p. 14

Libération, 1990. 'Pourquoi Vaulx-en-Velin?', 13–14 October

L'Express, 1991. 'Banlieue, immigration: l'état d'urgence', 5–12 June

Le Monde, 1991. 'Trop', and 'Le maire de Paris: "Il y a overdose"', 21 June, pp. 1 and 40

Le Monde, 1991. 'L'immigration et les mots', 22 June, pp. 1 and 8

Le Monde, 1992. 'La loi d'orientation sur la ville est en panne', 21 July, p. 9

Le Figaro, 1992. 'Banlieues: un rapport alarmant', 18 November, pp. 1 and 9

Le Monde, 1993. 'M. Pasqua: contre une société "pluriculturelle"', 21–2 March, p. 11

Le Figaro, 1993a. 'Pasqua veut rétablir les contrôles d'identité', 27 April, pp. 1 and 10

Le Figaro, 1993b. 'Le tonneau des Danaïdes' (Alain Peyrefitte), 27 April, p. 1

Le Figaro, 1993. 'Banlieues: les solutions du gouvernement', 28 April, pp. 1 and 6

Le Figaro, 1994. 'Le retour des casseurs', 11 March, pp. 1 and 10

Le Monde, 1995. 'M. Raoult veut légaliser le déplacement de familles indésirables', 20 July, p. 6

Le Monde, 1996. 'Le plan pour les banlieues prévoit la création de 100 000 emplois réservés aux 18–25 ans', 19 January, pp. 1, 8–9

Libération, 1996. '"Il faut en finir avec l'approche caritative des banlieues"', 19 January, p. 6

Le Monde, 1996. 'Trente-huit quartiers sensibles pressentis pour devenir "zones franches urbaines"', 30 March, p. 10

Le Monde, 1996. 'La part de la France' (Michel Rocard), 24 August, pp. 1 and 8

Le Monde, 1997. 'Le frigidaire de M. Debré et la baignoire de M. Le Pen', 30 April, p. 8

Le Monde, 1998. 'Le Front national, c'est ça', 21 March, pp. 14–15

Libération, 1998. 'Le procès de la fusillade Nation-Vincennes', 30 September, p. 15

Le Monde, 1998. 'La violence urbaine tend de plus en plus à toucher de nouvelles villes' (interview with Lucienne Bui-Trong), 8 December, p. 17

Le Monde, 2001. 'A Marseille, les anciens de l'immigration, oubliés de la rénovation urbaine', 22 June, p. 11

Libération, 2001. 'Le gouvernement s'attaque aux ghettos', 2 October, p. 21

Libération, 2001. 'Le lepénisme surnage dans le Beaujolais', 28 December, pp. 10–11

Libération, 2002. 'Vertus et défauts des recettes lyonnaises', 29 January, p. 3

Libération, 2002. 'Un collectif pour calmer le Val-Fourré', 11 February, p. 17

Libération, 2002. 'Violence: l'appel désespéré d'Isabelle', 15 February, p. 19

Le Monde, 2002a. 'Jacques Chirac engagé par ses partisans à "muscler" son discours sur la sécurité', 27 February, p. 10

Le Monde, 2002b. 'A Paris, place des Fêtes, "il suffit à certains de voir un groupe de jeunes pour se sentir agressés"', 27 February, p. 12

Libération, 2002. 'Violences policières à Paris', 28 February, p. 15

Le Monde, 2002. 'La police face à l'insaisissable économie souterraine des quartiers', 1 March, p. 9

Libération, 2002. 'Les Français ont-ils peur de tout?', 8 April, pp. 4–5

Le Monde, 2002. 'Les contrôles d'identité abusifs aggravent les tensions dans les cités', 20 April, p. 11

Le Monde, 2002. 'Sécurité: l'offensive Sarkozy', 18 May

Le Monde, 2002. 'Jean-Louis Borloo esquisse son plan pour une "nouvelle bataille de France"', 28 May, p. 12

Le Monde, 2002a. 'Le ministre veut publier plus fréquemment les chiffres de la délinquance', 31 May

Le Monde, 2002b. 'Sarkozy: ma politique de sécurité', 31 May

Libération, 2002. 'Chère répression', 11 July, pp. 1–3

Libération, 2002. 'Quartiers: Borloo promet des lendemains qui chantent', 3 October

Le Monde, 2003. 'Les zones franches urbaines créent-elles vraiment des emplois?', 18 February

Le Canard enchaîné, 2003. 'Une police de proximité à la sauce Sarko', 5 March, p. 4

Le Canard enchaîné, 2003. 'Comment la police d'en bas mitonne les statistiques d'en haut', 23 April, p. 4

Le Monde, 2003. 'Le gouvernement prévoit la démolition de 40 000 logements par an', 19 June

Libération, 2003. 'Des quotas de gardes à vue dans les commissariats de l'Hérault', 31 October

Le Monde, 2004. 'Les objectifs et priorités de Nicolas Sarkozy pour 2004', 14 January

Libération, 2004. 'Pas de quartier pour 20 quartiers', 15 January

Le Monde, 2004. 'M. Sarkozy veut restaurer "l'Etat de droit" dans les zones ciblées', 16 January

L'Humanité, 2004. 'Vingt-trois quartiers à l'index', 27 January

Le Monde, 2004a. 'Trois questions à Patrick Bruneteaux', 27 January

Le Monde, 2004b. 'Le nombre des violences policières en hausse en 2003 pour la sixième année consécutive', 27 January

Libération, 2004. 'Strasbourg: "On n'est pas du gibier"', 22 March

Libération, 2004. 'Un rapport sonne l'alarme sur les violences policières', 6 May

Libération, 2004. 'Le plus grand recul des droits de l'homme depuis l'Algérie', 25 May

Libération, 2005a. 'Politique de la ville: Borloo vend ses maisons', 24 February, pp. 6–7

Libération, 2005b. 'Des villes dans l'illusion de la démolition', 24 February, p. 7

Le Monde, 'Notre stratégie est la bonne' (Nicolas Sarkozy), 6 November

Libération, 2005. 'Banlieues: trois mesures phares', 8 November

Le Monde, 2005. 'La grenade lacrymogène des policiers ne visait pas la mosquée de Clichy-sous-Bois', 10 November, p. 12

Le Monde, 2005. 'Nicolas Sarkozy veut expulser les étrangers impliqués dans les violences urbaines', 11 November, p. 12

Libération, 2005. 'Sarkozy, à droite dans ses bottes', 21 November, p. 13

Le Monde, 2005. 'M. Sarkozy durcit son discours sur les banlieues', 22 November, p. 12

Libération, 2005. 'Le rapport qui contredit Sarkozy', 8 December, p. 16

Le Monde, 2005. 'Le gouvernement veut relancer les zones franches urbaines', 2 December, p. 14

Le Canard enchaîné, 2005. 'Plus de la moitié du plan sécuritaire de Le Pen déjà cannibalisé par la droite',14 December, p. 4

Libération, 2005. 'Muhittin, rescapé de Clichy, donne sa version', 16 December

Libération, 2006. 'Loi SRU: des sanctions plus lourdes pour les riches égoïstes', 7–8 January, p. 14

Le Canard enchaîné, 2006. 'L'abbé Pierre au karcher', 1 February, p. 5

Libération, 2006. '85 quartiers "sensibles" concernés', 8 May

Le Canard enchaîné, 2006. 'Les poulets au gril d'Amnesty', 24 May, p. 5

Le Monde 2006. 'Nouveaux incidents à Montfermeil et Clichy-sous-Bois, en Seine-Saint-Denis, 31 May

Le Canard enchaîné, 2006. 'Quand les RG cognent sur la police', 7 June, p. 4

Libération, 2006. 'Comment la "loi HLM" a survécu à la droite', 13 July

SECONDARY SOURCES

Aballéa F., 2000. 'Genèse d'une politique de la ville ou la ville comme catégorie de l'action publique', *FORS Recherche Sociale* 154, avril–juin: 4–21

Agora, 1995. *Pour une régie de quartier: Pré-projet soumis aux habitants* (pamphlet)

Amnesty International, 2005. 'France, the search for justice: The effective impunity of law enforcement officers in cases of shootings, deaths in custody or torture and ill-treatment', available at: http://web.amnesty.org/library/print/ENGEUR210012005 (last accessed 2 January 2007)

Appleton, A., 1999., 'The new social movement phenomenon: Placing France in contemporary perspective', *West European Politics* 22(4): 57–75

Bachmann, C. and Le Guennec, N., 1996. *Violences urbaines. Ascension et chute des classes moyennes à travers cinquante ans de politique de la ville* (Albin Michel, Paris)

Balibar, E., 2001. *Nous, citoyens d'Europe? Les frontières, l'Etat, le peuple* (La Découverte, Paris)

Banlieues 89, 1989. *Vers une civilisation urbaine. Assises de Nanterre 20 et 21 mai 1989* (Délégation Interministérielle à la Ville et au Développement Social Urbain and Banlieues 89, Paris)

Barthélémy, A., 1995. *Un avenir pour la ville. Face à la crise urbaine* (Esprit, Paris)

Battegay, A. and Boubeker, A., 1991–2. 'Des Minguettes à Vaulx-en-Velin: fractures sociales et discours publics', *Les Temps modernes* 545–6, décembre-janvier: 51–76

Begag, A., 1990. 'La révolte des lascars contre l'oubli à Vaulx-en-Velin', *Les Annales de la recherche urbaine* 49: 114–21

Béhar, D., 1998. 'Question urbaine et question sociale: quel lien pour quelle politique publique?', *Problèmes économiques* 2(574): 1–5

Béhar, D., 1999. 'En finir avec la politique de la ville?', *Esprit* 258: 209–18

Béhar, D., 2001. 'La politique de la ville: une politique a-territoriale?', *FORS Recherche Sociale* 158, avril–juin: 4–12

Belbahri, A., 1984. 'Les Minguettes ou la surlocalisation du social', *Espaces et sociétés* 45: 101–8

Blatt, D., 1997. 'Immigrant politics in a republican nation', in A.G. Hargreaves and M. McKinney (eds), *Post-Colonial Cultures in France* (Routledge, London), pp. 40–55

Bonelli, L., 2001. 'Renseignements généraux et violences urbaines' *Actes de la recherche en sciences sociales* 136–7, mars: 95–103

Bonelli, L., 2003. 'Une vision policière de la société', *Manière de voir, Le Monde diplomatique* 71: 38–41

Borloo, J.-L., 2004. 'Donner plus à ceux qui ont moins', *Problèmes politiques et sociaux* 906: 105

Boubeker, A., 1997. 'Vaulx-en-Velin dans la guerre des images: Les événements d'octobre 1990 et l'expérience de la visibilité publique', in J. Métral (ed.), *Les aléas du lien social: Constructions identitaires et culturelles dans la ville* (La Documentation française, Paris), pp. 87–101

Bourdieu, P., 1999. 'Site effects', in P. Bourdieu et al. (eds), *The Weight of the World: Social Suffering in Contemporary Society*, trans. by P. P. Ferguson, S. Emanuel, J. Johnson and S.T. Waryn (Stanford University Press, Stanford), pp. 123–9

Bourmeau, S., 2003. 'De la démocratie à la démagogie', *Les Inrockuptibles* 376: 30–5

Brenner, N. and Theodore, N., 2002. 'Cities and the geographies of "actually existing neoliberalism"', *Antipode* 34(3): 349–79

Brown, W., 2003. 'Neo-liberalism and the end of liberal democracy', *Theory & Event* 7(1) (online)

Brubaker, R., 1992. *Citizenship and Nationhood in France and Germany* (Harvard University Press, Cambridge, MA)

Brun, M., 2002, 'La loi d'orientation et de programmation pour la sécurité intérieure', *Regards sur l'actualité* 284: 5–12

Budgen, S., 2002. 'The French fiasco', *New Left Review* 17, September: 31–50

Buechler, S., 2000. *Social Movements in Advanced Capitalism: The Political Economy and Cultural Construction of Social Activism* (Oxford University Press, Oxford)

Bui-Trong, L., 1993. 'L'insécurité des quartiers sensibles: une échelle d'évaluation', *Les Cahiers de la sécurité interieure* 14, août–octobre: 235–47

Bui-Trong, L., 1998a. 'Les violences urbaines à l'échelle des renseignements généraux: un état des lieux pour 1998', *Les Cahiers de la sécurité interieure* 33: 215–24

Bui-Trong, L., 1998b. 'Sur quelques secrets de fabrication . . . Entretien avec Lucienne Bui-Trong', *Les Cahiers de la sécurité interieure* 33: 225–33

Bui-Trong, L., 2000. 'Violence urbaine dans les quartiers sensibles', in M.-F. Mattei and D. Pumain (eds), *Données Urbaines 3* (Anthropos, Paris), pp. 123–36

Bui-Trong, L., 2003. *Les racines de la violence: de l'émeute au communautarisme* (Louis Audibert, Paris)

Cahiers de la sécurité intérieure (Les), 2000 'Jeunes sans foi ni loi? Retour sur la délinquance des mineurs' (whole issue), no. 42

Campbell, D., 1998 [1992]. *Writing Security: United States Foreign Policy and the Politics of Identity* (University of Minnesota Press, Minneapolis)

Carrère, G., 1994. 'Ambitions et réalités du développement social urbain dans le Rhône dans les années 80–90', *Revue française d'administration publique* 71, juillet–septembre: 389–94

Cesari, J., 1997. *Faut-il avoir peur de l'Islam?* (Presses de Sciences Po, Paris)

Chabanet, D., 1999. 'La politique de la ville au défi de la participation des habitants à Vaulx-en-Velin', in R Balme, A. Faure and A. Mabileau (eds), *Les nouvelles politiques locales: dynamiques de l'action publique* (Presses de Sciences Po, Paris), pp. 345–64

Chaline, C., 1998 [1997]. *Les politiques de la ville*, second edition (Presses Universitaires de France, Paris)

Champion, J.-B. and Marpsat, M., 1996. 'La diversité des quartiers prioritaires: un défi pour la politique de la ville', *Economie et Statistique* 294–5: 47–65

Champion, J.-B., Goldberger, M.-F. and Marpsat, M., 1993. 'Les quartiers "en convention"', *Regards sur l'actualité* 196: 19–28

Charrier, M., 1991–2. 'Vaulx-en-Velin', *Les Temps modernes* 545–6, décembre–janvier: 92–9

Claris. Le Bulletin, 2002. 'Agir pour clarifier le débat sur l'"insécurité"', no. 1, avril

Cochrane, D.A., 2000. 'The social construction of urban policy', in G. Bridge and S. Watson (eds), *A Companion to the City* (Blackwell, Oxford), pp. 531–42

Collovald, A., 2000. 'Violence et délinquance dans la presse: Politisation d'un malaise social et technicisation de son traitement', in F. Bailleau and C. Gorgeon (eds), *Prévention et sécurité. Vers un nouvel ordre social?* (DIV, Saint-Denis La Plaine), pp. 39–53

Collovald, A., 2001. 'Des désordres sociaux à la violence urbaine', *Actes de la recherche en sciences sociales* 136–7, mars: 104–13

Connolly, W.E., 1991. *Identity/Difference. Democratic Negotiations of Political Paradox* (Cornell University Press, Ithaca and London)

Corrigan, P. and Sayer, D., 1985. *The Great Arch: English State Formation as Cultural Revolution* (Basil Blackwell, Oxford)

Cubero, J., 2002–3. 'Vingt ans de la politique de la ville', *Sciences Humaines* 39: 20–3

Cyran, O., 2003. 'Violences policières impunies', *Manière de voir, Le Monde diplomatique* 71: 30–3

Damamme, D. and Jobert, B., 1995. 'La politique de la ville ou l'injonction contra-dictoire en politique', *Revue française de science politique* 45(1): 3–30

Daoud, Z., 1993. 'Brève histoire de la politique de la ville', *Panoramiques* 12: 136–9

de Rudder, V., 1992. 'Immigrant housing and integration in French cities', in D.L. Horowitz and G. Noiriel (eds), *Immigrants in Two Democracies: French and American Experience* (New York University Press, New York and London), pp. 247–67

de Rudder, V., 2001. 'Politiques d'"immigration" en Europe. Du principe d'hos-pitalité à la règle d'inhospitalité', *VEI Enjeux* 125, juin: 24–33

de Rudder, V. and Poiret, C., 1999. 'Affirmative action et "discrimination justi-fiée": vers un universalisme en acte', in Philippe Dewitte (ed.), *Immigration et intégration: l'état des savoirs* (La Découverte, Paris), pp. 397–406

Deltombe, T., 2005. *L'islam imaginaire: la construction médiatique de l'islamophobie en France, 1975–2005* (La Découverte, Paris)

Depincé, K., 2003, 'La loi d'orientation et de programmation pour la ville et la rénovation urbaine', *Regards sur l'actualité* 296: 25–35

Dikeç, M., 2001., 'Justice and the spatial imagination', *Environment and Planning A* 33(10): 1785–1805

Dikeç, M., 2005. 'Space, politics, and the political', *Environment and Planning D: Society and Space* 23(2): 171–88

Donzelot, J., 2006. *Quand la ville se défait: quelle politique face à la crise des banlieues?* (Seuil, Paris)

Donzelot, J. and Estèbe, P., 1999. 'Réévaluer la politique de la ville', in R Balme, A. Faure and A. Mabileau (eds), *Les nouvelles politiques locales: dynamiques de l'action publique* (Presses de Sciences Po, Paris), pp. 321–44

Dubet, F., 1987. *La galère: jeunes en survie* (Fayard, Paris)

Dubet, F. and Jazouli A., 1984. 'Une nouvelle politique de prévention? Le cas de l'opération été 1982', *Revue internationale d'action communautaire* 11: 8.

Dubet, F. and Lapeyronnie, D., 1992. *Les quartiers d'exil* (Seuil, Paris)

Duchesne, S., 2005. 'Identities, nationalism, citizenship and republican ideology', in A. Cole, P. Le Galès and J. Levy (eds), *Developments in French Politics* (Pal-grave, New York), pp. 230–44

Ebel, E., 1999. *Les préfets et le maintien de l'ordre public en France au XIXe siècle* (La Documentation française, Paris)

Elstob, K., 1997–8. 'Hate (*La Haine*)', *Film Quarterly* 51(2), Winter: 44–9

Epstein, R., 2004. 'Disparition ou consolidation de la politique de la ville?', *Pro-blèmes politiques et sociaux* 906: 86–90

Epstein, R. and Kirszbaum, T., 2006. 'Après les émeutes, comment débattre de la politique de la ville?', *Regards sur l'actualité* 319: 39–48

Essassi, B., 1992. *Vaulx-en-Velin: rapport final* (Banlieuescopies, Paris)

Estèbe, P., 2001. 'Instruments et fondements de la géographie prioritaire de la politique de la ville (1982–1996)', *Revue française des affaires sociales* 3: 25–38

Estèbe, P., 2004. *L'usage des quartiers. Action publique et géographie dans la politique de la ville (1982–1999)*. Paris: L'Harmattan

Esterle-Hedibel, M., 2002. 'Jeunes des cités, police et désordres urbains', in L. Mucchielli and P. Robert (eds), *Crime et sécurité. L'état des savoirs* (La Décou-verte, Paris), pp. 376–85

Favier, G. and Kassovitz, M., 1995. *Jusqu'ici tout va bien: scénario et photographie autour du film LA HAINE* (Actes Sud, Arles)

Feldblum, M., 1994. 'Commentary: Reconsidering the "republican" model', in W.A. Cornelius, T. Tsuda, P.L. Martin and J.F. Hollifield (eds), *Controlling Immigration: A Global Perspective* (Stanford University Press, Stanford), pp. 177–9

Feldblum, M., 1999. *Reconstructing Citizenship: The Politics of Nationality Reform and Immigration in Contemporary France* (State University of New York Press, New York)

Foucault, M., 1977 *Discipline and Punish: The Birth of the Prison*, trans. by A Sheridan (Vintage Books, New York)

Foucault, M., 1984. 'Space, knowledge, and power', in P. Rabinow (ed.), *The Foucault Reader* (Penguin Books, Harmondsworth), pp. 239–56

Foucault, M., 1991 [1978]. 'Governmentality', in G. Burchell, C. Gordon and P. Miller (Eds), *The Foucault Effect: Studies in Governmentality* (University of Chicago Press, Chicago), pp. 87–104.

Germa, A., 2005. 'State violence in Clichy-sous-Bois: An eyewitness account. Clichy-sous-Bois: Lawlessness or injustice?', available at: http://www.kersplebedeb.com.2005riots/lmsi_germa.html (last accessed 2 January 2007)

GISTI 2004. *La réforme de la double peine: les mesures transitoires* (GISTI, Paris)

Goze, M., 2002, 'La loi Solidarité et renouvellement urbains, composante de la réforme territoriale', in F. Cuillier (ed.), *Les débats sur la ville* 4 (Editions Confluences, Bordeaux), pp. 15–37

Harburger, S., 1994. 'L'Etat face au malaise urbain au début des années 80', *Revue française d'administration publique* 71, juillet–septembre: 385–8

Hargreaves A.G., 1995. *Immigration, 'Race' and Ethnicity in Contemporary France* (Routledge, London and New York)

Hargreaves A.G., 1996. 'A deviant construction: The French media and the "Banlieues"', *New Community* 22(4): 607–18

Hargreaves A.G., 1997, 'Multiculturalism', in C. Flood and L. Bell (eds), *Political Ideologies in Contemporary France* (Pinter, London and Washington), pp. 180–99

Hargreaves, A.G. and McKinney, M. (eds) 1997. *Post-Colonial Cultures in France* (Routledge, London)

Harvey, D., 1989. *The Urban Experience* (Johns Hopkins University Press, Baltimore)

Holston, J., 1998. 'Spaces of insurgent citizenship', in L. Sanderock (ed.), *Making the Invisible Visible: A Multicultural Planning History* (University of California Press, Berkeley), pp. 37–56

Hubbard, P., 2004. 'Revenge and injustice in the neoliberal city: Uncovering masculinist agendas', *Antipode* 36(4): 665–86

Hunt, M.C. and Chandler, J.A., 1993, 'France', in J.A. Chandler (ed.), *Local Government in Liberal Democracies: An Introductory Survey* (Routledge, London), pp. 53–72

Idrac, F., 1996. 'La pacte de relance pour la ville', *Regards sur l'actualité* 222: 19–34

Jaillet, M.-C., 2000, 'La politique de la ville, une politique incertaine', *Regards sur l'actualité* 260: 29–45

Jaillet, M.-C., 2003. 'La politique de la ville en France: histoire et bilan', *Regards sur l'actualité* 296: 5–23

Jazouli, A., 1992. *Les années banlieues* (Seuil, Paris)

Jean, J.-P., 2003. 'La France à l'heure de la répression', *Manière de voir, Le Monde diplomatique* 71: 27–9

Jean, J.-P., 2004. 'Dix ans de réformes pénales: une recomposition du système judiciaire', *Regards sur l'actualité* 300: 33–48

Jennings, J., 2000. 'Citizenship, republicanism and multiculturalism in contemporary France', *British Journal of Political Science* 30(4): 575–97

Jobert, B. and Théret, B., 1994. 'France: La consécration républicaine du néo-libéralisme', in B. Jobert (ed.), *Le tournant néo-libéral en Europe. Idées et recettes dans les pratiques gouvernementales* (L'Harmattan, Paris), pp. 21–85

Jones, M. and Ward, K., 2002. 'Excavating the logic of British urban policy: Neo-liberalism as the "crisis of crisis-management"', *Antipode* 34(3): 473–94

Keith, M., 1993. *Race, Riots and Policing: Lore and Disorder in a Multi-racist Society* (UCL Press, London)

Knapp, A. and Wright, V., 2001. *The Government and Politics of France*, fifth edition. London: Routledge

Lagrange, H., 2006a. 'Autopsie d'une vague d'émeutes', in H. Lagrange and M. Oberti (eds), *Emeutes urbaines et protestations: une singularité française* (Sciences Po, Paris), pp. 37–58

Lagrange, H., 2006b. 'La structure et l'accident', in H. Lagrange and M. Oberti (eds), *Emeutes urbaines et protestations: une singularité française* (Sciences Po. Paris), pp. 105–30

Lagrange, H. and Oberti, M. (eds) 2006. *Emeutes urbaines et protestations: une singularité française* (Sciences Po, Paris)

Lapeyronnie, D., 1995. 'La politique de la ville et la représentation des populations', *Problèmes économiques* 2(418): 14–16

Larner, W., 2000. 'Neo-liberalism: Policy, ideology, governmentality', *Studies in Political Economy* 63 (Autumn): 5–25

Larner, W., 2003. 'Neoliberalism?', *Environment and Planning D: Society and Space* 21: 509–12

Le Galès, P., 1995. 'Politique de la ville en France et en Grande-Bretagne: volontarisme et ambiguïtés de l'Etat', *Sociologie du travail* 2: 249–75

Le Galès, P., 2005. 'Reshaping the state? Administrative and decentralization reforms', in A. Cole, P. Le Galès and J. Levy (eds), *Developments in French Politics* (Palgrave, New York), pp. 122–37

Le Galès, P. and Mawson, J., 1994. *Management Innovations in Urban Policy: Lessons from France* (The Local Government Management Board, London)

Le Goaziou, V. and Mucchielli, L. (eds) 2006. *Quand les banlieues brûlent . . . retour sur les émeutes de novembre 2005* (La Découverte, Paris)

Leclerc, H., 2004. 'Libertés publiques: l'année horrible', in Ligue des droits de l'Homme, *L'état des droits de l'Homme en France* (La Découverte, Paris), pp. 27–34

Lefebvre, H., 1970. *La révolution urbaine* (Gallimard, Paris)

Lefebvre, H., 1978. *De l'Etat IV: les contradictions de l'Etat moderne* (Union Générale d'Editions, Paris)

Lefebvre, H., 1996. *Writings on Cities*, trans. and introduced by E. Kofman and E. Lebas (Blackwell, Oxford)

Lemke, T., 2001. 'The birth of bio-politics: Michel Foucault's lecture at the Collège de France on neo-liberal governmentality', *Economy and Society* 30: 190–207.

Lenoir, R., 1974. *Les exclus: un Français sur dix* (Le Seuil, Paris)

Levy, J., 1999 *Tocqueville's Revenge: State, Society, and Economy in Contemporary France* (Harvard University Press, Cambridge, MA)

Levy, J., 2001. 'Partisan politics and welfare adjustment: The case of France', *Journal of European Public Policy* 8(2): 265–85

Levy, J., 2002. 'The state after statism: French economic and social policy in the age of globalization', paper presented at the Thirteenth International Conference of Europeanists, Chicago, 14–16 March

Levy, J., 2005. 'Economic policy and policy-making', in A. Cole, P. Le Galès and J. Levy (eds), *Developments in French Politics* (Palgrave, New York), pp. 170–94

Lochak, D., 2004. 'La loi sur la maîtrise de l'immigration: analyse critique', *Regards sur l'actualité* 299: 13–28

Loubière, A., 1984. 'Banlieues 89: piège à media ou politique?', *Architectes architecture* 153, décembre: 12–13

McCarthy, J. and Prudham, S., 2004. 'Neoliberal nature and the nature of neoliberalism', *Geoforum* 35: 275–83

Macé, E., 2002. 'Le traitement médiatique de la sécurité', in L. Mucchielli and P. Robert (eds), *Crime et sécurité. L'état des savoirs* (La Découverte, Paris), pp. 33–41

MacLeod, G., 2002. 'From urban entrepreneurialism to a "revanchist city"? On the spatial injustices of Glasgow's renaissance', *Antipode* 34(3): 602–24

Marchands de sécurité 2002. (Program transcript), directed by Richard Vargas, produced by Coup d'œil productions

Mazey, S., 1993. 'Developments at the French meso level: Modernizing the French state', in L.J. Sharpe (ed.), *The Rise of Meso Government in Europe* (Sage, London), pp. 61–89

Merlin, P., 1998. *Les banlieues des villes françaises* (La Documentation française, Paris)

Merlin, P. and Choay, F. (eds), 2000. *Dictionnaire de l'urbanisme et de l'aménagement.* (Presses Universitaires de France, Paris)

Mucchielli, L., 2001. *Violences et insécurité. Fantasmes et réalités dans le débat français* (La Découverte, Paris)

Murdoch, J., 2004. 'Putting discourse in its place: Planning, sustainability and the urban capacity study', *Area* 36: 50–8

Nakano, K., 2000. 'The role of ideology and elite networks in the decentralization reforms in 1980s France', *West European Politics* 23(3): 97–114

Nicholls, W., 2003. 'Poverty regimes and the constraints on urban democratic politics: Lessons from Toulouse, France', *European Urban and Regional Studies* 19(4): 359–72

Nuttens, J.-D., 2004. 'La loi portant adaptation de la justice aux évolutions de la criminalité', *Regards sur l'actualité* 300: 5–17

OECD, 1998. *Integrating Distressed Urban Areas* (OECD, Paris)

Panoramiques 1993. 'Banlieues: intégration ou explosion?', no. 12 (whole issue)

Parent, J.-F. and Schwartzbrod, J.-L., 1995. *Deux hommes, une ville: Paul Mistral, Hubert Dubedout, Grenoble* (La Pensée sauvage, Grenoble)

Paugam, S., 1996. 'Introduction: La constitution d'un paradigme', in S. Paugam (ed.), *L'Exclusion: l'état des savoirs* (La Découverte, Paris), pp. 7–19

Peck, J., 2001. 'Neoliberalizing states: Thin policies/hard outcomes', *Progress in Human Geography* 25(3): 445–55

Peck. J., 2003,. 'Geography and public policy: Mapping the penal state', *Progress in Human Geography* 27(2): 222–32

Peck, J., 2004. 'Geography and public policy: Constructions of neoliberalism', *Progress in Human Geography* 28(3): 392–405

Peck, J. and Tickell, A., 2002. 'Neoliberalizing space', *Antipode* 34(3): 380–404

Percy-Smith, J., 2000. 'Introduction: The contours of social exclusion', in J. Percy-Smith (ed.), *Policy Responses to Social Exclusion: Towards Inclusion?* (Open University Press, Buckingham and Philadelphia), pp. 1–21

Preteceille, E., 1988. 'Decentralization in France: New citizenship or restructuring hegemony?' *European Journal of Political Research* 16: 409–24

Raco, M., 2003. 'Governmentality subject-building, and the discourses and practices of devolution in the UK', *Transactions of the Institute of British Geographers* 28: 75–95

Raco, M. and Imrie, R., 2000. 'Governmentality and rights and responsibilities in urban policy', *Environment and Planning A* 32(12): 2187–204

Rajsfus, M., 2002. *La police et la peine de mort* (L'Esprit frappeur, Paris)

Rancière, J., 1994. 'Post-democracy, politics and philosophy: An interview with Jacques Rancière', *Angelaki* 1(3): 171–8

Rancière, J., 1995. *La Mésentente: politique et philosophie* (Galilée, Paris).

Rancière, J., 1999 [1995]. *Dis-agreement: Politics and Philosophy*, trans. by J. Rose (University of Minnesota Press, Minneapolis)

Rancière, J., 2000a. *Le partage du sensible: esthétique et politique* (La Fabrique, Paris)

Rancière, J., 2000b. 'Biopolitique ou politique?' (interview with Jacques Rancière) *Multitudes*, available at: http://multitudes.samizdat.net/Biopolitique-ou-politique. html (last accessed 2 January 2007)

Rancière, J., 2003a. 'Politics and aesthetics: An interview', *Angelaki* 8(2): 191–211

Rancière, J., 2003b. 'The thinking of dissensus: Politics and aesthetics', paper presented at the conference *Fidelity to the Disagreement: Jacques Rancière and the Political*, Goldsmiths College, London, 16–17 September

Recherches 20 1991. 'Loi d'orientation pour la ville: séminaire chercheurs décideurs' (Ministère de l'équipement, des transports et du logement, Paris)

Rey, H., 1999. 'La peur des banlieues', in P. Dewitte (ed.), *Immigration et intégration: l'état des savoirs* (La Découverte, Paris), pp. 274–8

Robson, M., 2005. 'Introduction: Hearing voices', *Paragraph* 28(1): 1–12

Roché, S., 2005. *Police de proximité: nos politiques de sécurité* (Seuil, Paris)

Rose, N., 1996. 'Governing "advanced" liberal democracies', in A. Barry, T. Osbourne and N. Rose (eds), *Foucault and political reason* (UCL Press, London), pp. 37–64

Rose, N., 1999. *Powers of Freedom: Reframing Political Thought* (Cambridge University Press, Cambridge)

Rose, N. and Miller, P., 1992. 'Political power beyond the state: Problematics of government', *British Journal of Sociology* 43: 173–205

Safran, W., 1990. 'The French and their national identity: The quest for an elusive substance', *French Politics and Society* 8(1): 56–67

Schmidt, V.A., 2002 *The Futures of European Capitalism* (Oxford University Press: Oxford)

Schmidt, V.A., 2003. 'French capitalism transformed, yet still a third variety of capitalism', *Economy and Society* 32(4): 526–54

Shapiro, M. and Neubauer, D., 1989. 'Spatiality and policy discourse: Reading the global city', *Alternatives* 14(3): 301–25

Sharma, S. and Sharma, A., 2000. '"So far so good . . ." *La Haine* and the poetics of the everyday', *Theory, Culture & Society* 17(3): 103–16

Silver, H., 1993. 'National conceptions of the new urban poverty: Social structural change in Britain, France and the United States', *International Journal of Urban and Regional Research* 17(3): 336–54

Silver, H., 1994. 'Social exclusion and social solidarity: three paradigms', *International Labour Review* 133(5–6): 531–78

Silverman, M., 1992. *Deconstructing the Nation: Immigration, Racism and Citizenship in Modern France* (Routledge, London)

Simon, P., 1995. 'La politique de la ville contre la ségrégation. Ou l'idéal d'une ville sans divisions', *Les Annales de la recherche urbaine* 68–9: 27–33

Simon, P. and Lévy, J.-P., 2005. 'Questions sociologiques et politiques sur la "mixité sociale"', *ContreTemps* 13, mai: 83: 92

Simons, J., 1995. *Foucault and the Political* (Routledge, London)

Smith, R.W., 1987, 'Towards *autogestion* in socialist France? The impact of industrial relations reform', *West European Politics* 10(1): 46–61

Tchétché-Apéa, P.-D., 2000. 'Révéler la citoyenneté des banlieues', in H. Hatzfeld (ed.), *Banlieues: villes de demain, Vaulx-en-Velin au-delà de l'image* (Editions du CERTU, Lyon), pp. 73–90

Tévanian, P. and Tissot, S., 1998. *Mots à maux: Dictionnaire de la lepénisation des esprits* (Editions Dagorno, Paris)

Thrift, N., 1999. 'Cities and economic change: Global governance?', in J. Allen, D. Massey and M. Pryke (eds), *Unsettling Cities* (Routledge/Open University, London/Buckingham), pp. 271–308

Vaulx-en-Velin, ma ville . . . 1996 (Vaulx-en-Velin)

Wacquant, L., 1992. 'Pour en finir avec le mythe des "cités-ghettos": les différences entre la France et les États-Unis', *Annales de la recherche urbaine* 52: 20–30

Wacquant, L., 1993. 'Urban outcasts: Stigma and division in the black American ghetto and the French urban periphery', *International Journal of Urban and Regional Research* 17(3): 366–83

Wacquant, L., 1995. 'The comparative structure and experience of urban exclusion: "Race", class, and space in Chicago and Paris', in K. McFate, R. Lawson and W.J. Wilson (eds), *Poverty, Inequality, and the Future of Social Policy* (Russell Sage Foundation, New York), pp. 543–70

Wacquant, L., 1999. 'America as social dystopia: The politics of urban disintegration, or the French uses of the "American Model"', in P Bourdieu et al. (eds), *The Weight of the World: Social Suffering in Contemporary Society*, trans. by P.P. Ferguson, S. Emanuel, J. Johnson and S.T. Waryn (Stanford University Press, Stanford), pp. 130–9

Wacquant, L., 2001. 'The penalization of poverty and the rise of neo-liberalism', *European Journal on Criminal Policy and Research* 9: 401–12

Wacquant, L., 2003. 'Des contes sécuritaires venus d'Amérique', *Manière de voir, Le Monde diplomatique* 71: 10–15

Waters, S., 1998. 'New social movement politics in France: The rise of civic forms of mobilization', *West European Politics* 21(3): 170–86

Weber, S., 1992. 'In the name of the Law', in D. Cornell, M. Rosenfeld and D.G. Carlson (eds), *Deconstruction and the Possibility of Justice* (Routledge, New York), pp. 232–57.

Weil, P., 1991. *La France et ses étrangers: l'aventure d'une politique de l'immigration, 1938–1991* (Calmann-Lévy, Paris)

Weil, P. and Crowley, J., 1994. 'Integration in theory and practice: A comparison of France and Britain', *West European Politics* 17(2): 110–26

Wieviorka, M., 1998. 'Le multiculturalisme', *Cahier du CEVIPOF* 20: 74–89

Wieviorka, M. et al., 1999. *Violence en France* (Seuil, Paris)

Index

actually existing republicanism 175,
177
  paradox of 175, 177
affirmative action *à la française* 12,
99–103
  see also discrimination
*Agence Nationale pour la Rénovation
  Urbaine* (ANRU) 120–3, 156,
  191n, 192n
Algerian war 144, 175
Amnesty International 160–1
anti-ghetto law 12, 76
  see also Loi d'Orientation pour la Ville
  (LOV)
autogestion 37, 41, 49, 64, 66, 99, 179n
  see also self-management

Bachmann, Christian 39, 49
Balladur, Edouard 29, 78, 92, 93, 96
banlieues
  and difficulties of political
    mobilization 17, 131, 147,
    149–51
  and Islam 73–4, 89, 175–6, 183n
  and police violence 15, 90, 160–4,
    166, 169
  and sensible evidences 21, 80, 87,
    168
  and stigmatization 148, 154–5, 157,
    166, 175, 176

and terrorism 3, 4, 74, 87, 89, 164,
    167, 176
  as badlands 7, 15, 22, 126, 149
  as lawless 79, 144–5
  as sites of political mobilization 15,
    22, 126, 147, 149, 173
  as threat 4, 8, 11, 14, 15, 22, 31, 65,
    73, 91, 94, 95, 124, 126, 157,
    167, 169, 173, 175–6
  colonial government of 144–5
  definition of 7–8, 178n
  difference from suburbs 7–8
  Marshall Plan for 93, 97
  media construction of 8–9, 183n
  paternalistic government of
    144–5
  see also cités; working-class
    neighbourhoods
Banlieues 89 56, 60, 62
Béhar, Daniel 4, 101, 103
Besson Law (1990) 76, 77
beur 58, 143
  movement 145
bidonvilles 38
Bonelli, Laurent 82
Bonnemaison Report 50–1, 54
Borloo Law (2003) 120–4
  neoliberal orientation of 124
Brixton 7, 40, 42–4, 65, 174
Brown, Wendy 25, 27

Bui-Trong, Lucienne 81–3, 130,
   158–60, 167–9, 185n, 191n
   Bui-Trong scale 82, 87, 153
   see also French Intelligence Service

Cardo Report 80
Carrez Law (1995) 78
Chanteloup-les-Vignes 69, 70, 71, 79,
   80
   La Haine 182n
Chevènement, Jean-Pierre 106
   on little savages (sauvageons) 166
Chevènement Law (1999) 112–13
Chirac, Jacques 57, 94, 96
   2002 presidential elections 91
   2005 revolts 152
   and neoliberalism 29, 30, 99
   on immigrants' noise and smell 91
   on urban policy 60, 97, 98
   on zero tolerance 118
cités 39, 70, 79, 90, 93, 97–8, 135,
   146–7, 157, 161–2, 186–7n, 191n
   definition of 13–14
citizenship 62, 65, 69, 71, 78, 94, 130,
   134–5, 145–6
   insurgent 15, 149, 178n
   new 41
   republican idea of 10, 28
   urban 147, 151
City Contracts 63, 101, 106, 108,
   111–12, 121–3
Clichy-sous-Bois 153, 160–1
Cochrane, Allan 22
cohabitation 60, 62, 104
   definition of 182n
Comité Interministériel des Villes
   (CIV) 62, 110
Commission Nationale pour le
   Développement Social des Quartiers
   (CNDSQ) 48–50, 55, 62, 66,
   180n
communitarianism 90, 147, 151
   definition of 11
   see also ethnic separatism
Conseil National de Prévention de la
   Délinquance (CNPD) 62, 181n

Conseil National des Villes (CNV) 62,
   87
Contrat Première Embauche (CPE) 28,
   179n
Contrats Locaux de Sécurité
   (CLS) 107–8, 125, 163
Corrigan, Philip 5, 16, 22, 170
Courneuve, La 52, 55, 165
CRS see riot police

decentralization 29, 41, 47, 50, 59, 75,
   150
Delebarre, Michel 76, 78
Delarue, Jean-Marie 78
Delarue Report 78–9, 95
Délégation Interministérielle à la Ville
   (DIV) 62–3, 180n
democracy
   foot soldiers of 50
   local 49, 54, 150
Développement Social des Quartiers
   (DSQ) 50, 52–6, 60–3, 70, 88,
   92, 101, 111, 130, 156
Développement Social Urbain (DSU) 63,
   111
dirigiste state, dismantling of 29
Discipline and Punish (Foucault) 20
discrimination 15, 22, 89, 117, 135,
   152, 160, 166, 176–7
   Anglo-Saxon 100
   geographies of 169, 177
   positive 11–12, 51, 99–100, 104–6
   see also affirmative action à la
   française
double peine 89
   definition of 185–6n
Dubedout, Hubert 48–51, 56, 59, 62,
   64, 71, 75, 92, 157
   and autogestion 49, 99
   and Mitterrand 182n
Dubedout Report 50–4, 76, 101, 181n

economic restructuring 5, 33, 43, 53,
   117, 131–2, 154
emplois de ville 103
emplois jeunes 103, 192n

émeutes see revolts
Estèbe, Philippe 33, 60, 83, 99, 101,
    123, 124
ethnic separatism 11, 151, 175, 176
    see also communitarianism
ethnicity 10, 100–1, 177
European Union 176
    and social exclusion 63–4

Feldblum, Miriam 30, 57, 59, 94
Fifth Republic 42, 62, 175
Foucault, Michel
    and Rancière 19–20, 179n
    on governmentality 22–3
    on power and space 20
    on the police 19
French Intelligence Service
    (*Renseignements Généraux*, RG) 21,
    69, 73, 80–3, 85, 86, 89, 119, 125,
    147, 162, 164, 167, 172
    Cities and *Banlieues* section 81–2,
    158, 167, 185n, 191n
French Nationality Code 94
Front National (FN) 56–7, 60, 91,
    181n, 186n

Gaudin, Jean-Claude 97, 98, 99
Geindre Report 94–6
geographies of French urban policy 33
    local geography 60, 66, 83, 88
    relative geography 83, 85, 88, 101
    statist geography 88, 101, 111,
    122–3
ghettoization 75, 98
ghettos, connotations in France 11–12,
    138
Giscard d'Estaing, Valéry 40, 43
government, as spatial organization 18
governmentality 22–3, 25
grands ensembles 37–9, 40, 50, 51, 52,
    53, 62, 75, 96, 100, 103, 131, 132
Grands Projets de Ville (GPVs) 121, 138
Grands Projets Urbains (GPUs) 111
Groupes d'Action Municipale
    (GAMs) 49
Gulf War 73, 176

Habitat et Vie Sociale (HVS) 37, 38–9,
    180–1n
    critique of 53–4
Habitation à Loyer Modéré (HLM) 38,
    85, 91, 93, 113, 116, 131
    definition of 13
Haine, La (film) 159, 182n, 191n
Hargreaves, Alec 8, 10, 12, 38, 57, 73
Haut Conseil à l'Intégration (HCI) 11,
    79, 91–2, 183n, 184n
Holston, James 15, 149, 178n
hot summer of 1981 4, 39–40, 42, 46,
    51, 56, 72, 153, 176

identity checks 93, 94, 119, 145, 148,
    153, 161, 183n
immigrants 3, 12, 38, 51, 53, 54,
    57–8, 112, 116, 177
    as scapegoats 52
    as threat 59
    concentration of 12, 42, 51, 98
    expulsion of 40, 41, 42, 43, 175
    integration of 79, 114, 124, 175
    non-European 40, 79, 138, 175, 176
    North African 40, 73
    overdose of 91
    second-generation 50, 52, 58, 89,
    144
    voting rights for 41, 42, 49, 59, 144,
    176
immigration
    fear of 8
    problem of 175
    zero 94
insertion 50, 52, 53, 71, 77, 78, 100,
    120
*Institut National d'Etudes
    Démographiques* (INED) 168
*Institut National de la Statistique et des
    Etudes Economiques* (INSEE) 80,
    83, 87, 101
    City section 83
integration 79, 87, 94, 158
    lack of 108
    of foreigners 106
    of neighbourhoods 98, 110

republican model of 11, 72, 79, 90, 124, 177
urban 96–7, 98
see also immigrants, integration of
Intifada 7, 174, 176
of the banlieues 73
Islam 73, 89, 175, 176
and 2005 revolts 176
and banlieues 74, 183n
fear of 183n
Islamic fundamentalism 175, 176
Islamic headscarf affair 7, 73, 79, 174, 176
explained 183n
Islamphobia 183n

Jennings, Jeremy 10, 12, 73
Jospin, Lionel 74, 103, 104, 106
2002 presidential elections 118
on security 107–8, 118
on urban policy 107, 109–11, 113–14
Juppé, Alain 28, 96, 97
Justice in Banlieues, programme 89

Lagrange, Hugues 156, 157
laïcité see secularism
Larner, Wendy 25
Le Galès, Patrick 62, 113
Le Guennec, Nicole 39, 49
Le Pen, Jean-Marie 56, 57, 91, 119, 189n
League of Human Rights 120
Lefebvre, Henri 37
influence on urban policy 181n
on autogestion 179n
on the right to the city 77
Levy, Jonah 29, 104
Levy Report 61–2, 79
Loi d'Orientation pour la Ville (LOV) 12, 76–7, 99, 103, 106, 111, 112, 113, 136, 184n
Loi de Solidarité et Renouvellement Urbains (SRU) 112–13, 123
Loi pour la Sécurité Intérieure (LSI) 119

Loi relative à la Sécurité Quotidienne (LSQ) 118–19
Los Angeles riots 7, 73, 95, 174
LOV see Loi d'Orientation pour la Ville
see also anti-ghetto law
LSI see Loi pour la Sécurité Intérieure
LSQ see Loi relative à la Sécurité Quotidienne
Lyon 15, 39, 42, 43, 48, 51, 58, 75, 112, 130–4, 136, 147, 148, 156, 161

Maison de Justice et du Droit (MJD) 80–1, 106, 107, 108, 125
Mantes-la-Jolie 79, 115, 116, 154
Marchands de sécurité (2002) 163
Marche pour l'égalité et contre le racisme 58, 143, 180n
Marseille 48, 52, 115, 156
Merlin, Pierre 38, 39, 56, 77
MIB see Mouvement de l'Immigration et des Banlieues
Minguettes, Les 39, 42–3, 46–8, 51, 52, 58, 70, 72–3, 168
Ministry for the City 81–3, 87, 92
creation of 75–6
Ministry of Justice 14, 21, 69, 73, 80–1, 89, 125, 167, 168, 172
Ministry of the Interior 81, 96, 168
Mitterrand, François 41, 56, 58, 60, 62, 72
and Dubedout 182n
on urban policy (Bron speech) 75–6
on voting rights for immigrants 42, 58–9, 144
MJD see Maison de Justice et du Droit
Mouvement de l'Immigration et des Banlieues (MIB) 89–90, 115, 119, 139, 149–50, 151, 163–4, 166, 173, 185n
Mucchielli, Laurent 39

national identity, threats to 30
nationalism see republican nationalism
neoliberalism 6, 24–7
à la française 28–9

neoliberalism (cont'd)
and French republicanism 6, 31–2, 174
and state restructuring 25, 31, 174–5
as governmentality 25
as hybrid 6, 24, 27
as political rationality 24 –7, 28, 174
urban 26–7
neoliberalization 5, 24–6, 29–31, 33, 174
no-go areas, re-conquer 15
Nord-Pas-de-Calais 116

Opérations de Renouvellement Urbain (ORUs) 121
outlaw areas 93, 94, 107, 120, 164, 169

Pacte de Relance pour la Ville 12, 97–105, 111, 116, 121, 122, 123, 124
as Marshall Plan for banlieues 97
meaning of 186n
neoliberal orientation of 102–3
Parti Communiste Français (PCF) 41, 42, 57
Parti Socialiste (PS) 41
see also Socialists
participation of inhabitants 39, 54, 95, 110, 140, 146, 150
as eternal leitmotiv of urban policy 75, 123
participation simulacra 139
partition of the sensible see Rancière, Jacques
Pasqua, Charles 93, 96
on multi-cultural society 30, 92
Pasqua Law (1995) 99–100
Pasqua–Méhaignerie Laws (1993) 94
PCF see Parti Communiste Français
Peck, Jamie 27, 32
penal state 6, 14, 125–6, 167
European 32
geographies of 33

see also republican penal state; urban policy
Perben Law I (2002) 119
Perben Law II (2004) 119
Pesce Report 76
Peyrefitte Report 40
police, the
and politics 21, 173
and sensible evidences 87, 172
as inherently spatial 20
as order of governance 131
as organization and government 18
as regime of representation 18–19
as spatial ordering 20
spaces of 21, 22, 173
police de proximité 108, 120, 125, 164, 165, 187–8n
police impunity 161
Amnesty International on 160–1
police violence 15, 90, 152, 160–4, 166
geographies of 169
policy see urban policy
political mobilization see banlieues
political numbers, Rose on 86–7
political rationalities 5, 24, 174
see also neoliberalism
political spaces
constitution of 177
opening up 22, 147, 151
politics
and aesthetics 20
and power 20
and the police 21, 173
as disruption 18
as inherently spatial 21
Rancière on 17–21
spaces of 21, 173
politics of space, French urban policy 173
politique de la ville, la 4, 114, 150
positive discrimination see affirmative action à la française; discrimination
PS see Parti Socialiste

quotas 99

racism 15, 52, 59, 91, 106
Rancière, Jacques 5, 16–21, 23, 80,
    131, 172
  on Foucault 19–20, 179n
  on politics 17–21
  on politics and the police 20–1
  on sensible evidences 18–19
  on the partition of the
    sensible 18–20
  on the police 18–20
Rassemblement pour la République
    (RPR) 29, 91, 93, 179n
  local alliances with the FN 57, 60
Renseignements Généraux (RG) *see*
    French Intelligence Service
repression 15, 17, 23, 32, 33, 120,
    126, 157, 158, 163, 164, 166,
    168–9, 175
  geographies of 158–62, 177
republic, the
  as one and indivisible 4, 6, 29, 32,
    138, 177
  conception of 10–12, 28
  haunting of 11, 40–3, 65, 73, 94–7
republican model of integration 11,
    72, 79, 90, 124, 177
republican nationalism 11, 31, 79,
    90
  and urban policy 11, 14, 79
republican penal state 21, 31–4, 126,
    174–5
  and urban policy 31–4
  see also penal state
republican rhetoric 30, 113
republican state, contradictions
    of 28–31
republicanism 7, 10, 12, 30–1, 174,
    177
  and neoliberalism 6, 31–2, 174
  see also actually existing
    republicanism
resistance 6, 90
  as spatial 7, 173
Resistance of *Banlieues* 90, 185n
Revenu Minimum d'Insertion (RMI) 64,
    111

revolts
  and police provocation 161–2
  and spatial injustice 155
  and state of emergency 34, 166, 175
  as unarticulated justice
    movements 152–69
  geographies of 153–8, 169
  political significance of 17, 21–2, 90,
    152–3, 168–9, 173, 176
  triggering incidents 47, 90, 130,
    152, 153, 158–60
  see also social unrest
révolution urbaine, La (Lefebvre) 37
Rey, Henri 8, 39
RG *see* French Intelligence Service
Rhône-Alpes 43, 51, 116, 132–4
right to difference 51, 91
right to housing 76
  see also Besson Law
right to security 107, 110, 114, 162
right to the city
  in urban policy 14, 34, 54, 56, 66,
    76–7, 157
  Lefebvre on 37, 77
  see also anti-ghetto-law; *Loi
    d'Orientation pour la Ville* (LOV)
riot police (CRS) 130, 159
  responsibility in revolts 161–2
riots *see* revolts
RMI *see* Revenu Minimum d'Insertion
Rocard, Michel 62, 69, 72, 87, 130,
    183n, 185n
Rose, Nikolas 23, 86–7
RPR *see* Rassemblement pour la
    République
Rudder, Véronique de 12, 77, 79
Rushdie affair *see* Salman Rushdie affair

Salman Rushdie affair 7, 73, 174, 176
Sardais Report 69–71
Sarkozy, Nicolas 19, 21, 113, 119,
    120, 164
  2005 revolts 119, 160, 165–6, 191n
  on repression 120
Sarkozy Law (2002) 119
Sartrouville 79

Sayer, Derek 5, 16, 22, 170
Schmidt, Vivienne 28, 29, 30
Schwartz Report 50–1, 54
secularism (*laïcité*) 12, 183n
self-management 14, 179n
  see also autogestion
sensible evidences *see* police, the;
    Rancière, Jacques; urban policy
sensitive neighbourhoods 80–2, 89, 90,
    106, 107, 120, 153, 158, 164, 167
  cartography of 87
Silver, Hilary 28, 29, 30, 113
social exclusion, as urban policy
    problem 63, 76
social mixity 11, 77, 99, 121, 124–5,
    136, 137, 148
  definition of 12
social policy 28–9
social unrest 33
  see also revolts
Socialists 31, 40, 41, 60, 107, 143,
    144, 175
  see also Parti Socialiste
SOS-Racisme 181n, 191n
  distrust towards 145–6
space
  conceptualizations of 5, 24, 33
  contestation for 5, 7, 89, 173
  practices of articulation 4, 16, 87,
    170, 172
spaces of politics 21, 173
  opening up 17, 173
spatial injustice 155
spatiality, of orders of governance 20
SRU *see Loi de Solidarité et
    Renouvellement Urbains*
state restructuring 31
  and neoliberalism 25, 31, 174–5
  and urban policy 4, 5–6, 24, 33,
    174–5
  France 10, 26, 29, 30–1, 174
state's statements 5, 16, 32, 89, 90,
    94, 96, 126, 138, 149, 151, 166,
    170–1, 173, 174
Sueur Report 104–6, 110, 111, 121
suburbs *see banlieues*

Tickell, Adam 27
trente glorieuses 38, 95

UDF *see Union pour la Démocratie
    Française*
UMP *see Union pour un Mouvement
    Populaire*
Union pour la Démocratie Française
    (UDF) 57
Union pour un Mouvement Populaire
    (UMP) 179n
urban neoliberalism 26–7
urban policy
  and governmentality 22–3
  and republican nationalism 11, 14,
    79
  and sensible evidences 6
  and space 22–4
  and state restructuring 4, 5–6, 24,
    33, 174–5
  and the republican penal state 31–4
  as place-making 5, 7, 22, 32–3, 89,
    151, 171
  as police order 33
  as regime of representation 5, 21,
    172
  as spatial arrangement 23
  divisive spatiality of 14
  governmentality approaches 5–6, 23,
    26
  political economy approaches 5–6,
    26
  social constructionist approaches
    5–6
  see also geographies of French urban
    policy
urban violence 87, 89, 90, 96, 98, 126,
    164, 167, 169
  and the French Intelligence
    Service 81–2, 125, 147
  entering the agenda 81
  statistics of 167–8
  see also Bui-Trong scale; French
    Intelligence Service

Val-Fourré 154, 159, 164

Vaulx-en-Velin 15, 17, 21, 39, 43, 58,
    69, 71–4, 75, 79, 80, 81, 87, 89,
    95, 102, 112, 115, 117, 129–51,
    153, 155–6, 158, 164, 173, 175
  citizens of 143, 146–7
Vénissieux 39, 42–3, 79, 115, 116,
    168
Villepin, Dominique de 179n, 192n
Villepinte colloquium 106–7, 125
Voynet Law (1999) 111–12, 113

Wacquant, Loïc 11, 32, 154
working-class neighbourhoods 14, 41,
    53, 65, 118, 125, 154, 157

youth employment programme 103

zero tolerance 118, 163–4
zone, la 3, 13, 105
Zone à Urbaniser par Priorité (ZUP) 13,
    38, 40, 50, 105, 131–3, 137–8
Zone d'Aménagement Concerté
    (ZAC) 38, 105
Zone d'Education Prioritaire (ZEP) 12,
    105, 181n
Zone de Redynamisation Urbaine
    (ZRU) 100, 105, 116
Zone Franche Urbaine (ZFU) 100, 102,
    104–5, 111, 116, 121–2, 156, 157
Zone Urbaine Sensible (ZUS) 99–100,
    105, 123, 124, 156
zones de non-droit see outlaw areas
zoning, French urban policy 104–5

Printed and bound by CPI Group (UK) Ltd, Croydon, CR0 4YY

27/10/2024

14580366-0003